SOUND HEALING

How to Use Sound to Beat Stress & Anxiety

Farzana Ali

WATKINS
Sharing Wisdom
Since 1893

Sound Healing
Farzana Ali

First published in the UK and USA in 2024 by
Watkins, an imprint of Watkins Media Limited
Unit 11, Shepperton House, 83–93 Shepperton Road
London N1 3DF

enquiries@watkinspublishing.com

Commissioning Editor: Lucy Carroll
Project Editor: Brittany Willis
Head of Design: Karen Smith
Design Concept: Sarah O'Flaherty
Illustrator: Alice Claire Coleman
Production: Uzma Taj

A CIP record for this book is available from the British Library

ISBN: 978-1-78678-764-4 (Paperback)
ISBN: 978-1-78678-765-1 (eBook)

10 9 8 7 6 5 4 3 2 1

Printed and bound by CPI Group (UK) Ltd
Typeset by Lapiz

www.watkinspublishing.com

Publisher's Note:
This book includes a QR code to an accompanying audio track. This is not suitable for use while driving, operating heavy machinery or when requiring peak concentration.

To my beloved grandparents for teaching me
that love is the greatest vibration of all.

And to my darling Isaac and precious Layla
– the sound of your giggles is my therapy.

CONTENTS

Introduction

"I'm so tired" – we use and hear this phrase all the time. Yet rather than seeing it as a call to action to make time for ourselves and engage in active rest, we wear it (unfortunately) as a badge of honour. And then we compete with everyone around us to work out who is the most tired, the most exhausted, the most knackered among us.

We have convinced ourselves that in the invisible hierarchy of stress and burnout, each of us is at the top. We are trapped in a hamster wheel of self-destruction, and I can understand why. Before becoming a sound therapist, I was playing the same game.

In my fast-paced corporate job, my days started at 6am and could very easily end at midnight. I lived in a one-bedroom apartment yet could manage for days without once walking into my living room/kitchen. I was constantly on the go, and there was simply no time to slow down. Everything was urgent and had to be finished, completed, done.

When I would stop (whether voluntarily or involuntarily by way of burnout), I was in the same pattern; even though I thought I was resting, I definitely was not. I went on fast-paced holidays where every moment was mapped out. I felt a need to squeeze every drop of excitement from my time off work, so I would take the red-eye flight home and arrive at the office straight from the airport with my suitcase in tow.

Then there were the weekends when the burnout would hit. This involved doom-scrolling on social media for hours and being asked, "Are you still watching?" by Netflix. On other occasions, I would arrive at a friend's house for dinner only to fall deeply asleep on their sofa. Fortunately, my friends were kind enough to tuck me up into their spare room or feed me a nutritious meal before putting me in a taxi home.

"It's okay, I'll rest this weekend," I would mouth to them from the back of a cab, while wearily waving a hand goodbye. "I'm fine. Things aren't as bad as they looked," I'd reassure them when they called to check up on me.

I was tired and exhausted. So, what did I do about it? For a long time, I did nothing. I pushed myself forward and carried on.

The problem for most of us, I believe, is that when we are that tired, we simply do not have the capacity to teach ourselves out of these harmful patterns of behaviour. We know that meditation is helpful, but we do not have the energy to learn it. It is a wonderful practice but can take years to fully master.

We know that yoga is also beneficial, but the thought of donning a Lycra two-piece and learning how to bend into any shape other than lying down in savasana can be daunting. Also, not everyone has the privilege of having the space and privacy to do yoga, and yoga studios can feel exclusionary and ableist (although this is gradually changing).

Sleep is crucial – ideal, in fact – to help you truly rest and recover. But with whizzing thoughts and a collective inability to "switch off", sleep is like your favourite restaurant that is closed. The delicious memories are there – but you can't visit it.

Running and other high-octane activities that increase your heart rate are good for optimal health over time. Yet when exhaustion and/or anxiety meet a pounding heart rate, it can often feel like too much for your already highly wired body.

Adding more into your life to combat your anxiety feels overwhelming, but slowing down to deal with stress feels impossible, too. Maybe even boring.

This reminds me of the time I went to visit an Ayurvedic practitioner. I was running late and rushing, primarily because I had squeezed more into my day than was humanly possible. When I eventually arrived, I burst into the meeting room, apologizing for my tardiness, and rushed to introduce myself without even stopping to catch my breath.

I spoke so quickly that the practitioner felt her own speech speed up. She asked if I ever practised yoga or meditation. I answered in the negative, declaring it to be "boring" and that I could not concentrate when doing it and found myself getting irritated at its glacial pace. She said that meditation is essential for those who find it irritating, as these are the people who need it the most. Her words had a profound impact on me.

I think that was the first time I really took stock of my life's pace and how little rest was in it.

STRESS OVERLOAD

That level of tiredness and exhaustion rarely comes without its bedfellow – stress. And, boy, was I stressed. The pressures of what, on reflection, really was an impossible job were exacerbated when one of my team members went on maternity leave. I was juggling so much already in a role that countless people had tried and failed. In fact, when I left after five years, someone pointed out how *nine* people had tried to do my job in the five years before me. Long before that conversation happened, things really came to a head when an important cog of my tightly run ship was taken out of the equation. I hit the iceberg and the water came rushing in.

One afternoon I was sitting in the staff canteen, frozen, unable to return to my desk, my limbs heavy with panic. It was as though my body, my brain, my entire being was saying *no*. There was no more "getting on with it", no more "just ploughing through". Thankfully I no longer experience this, but the vivid memory remains seared into my brain, stored as a reminder – a warning – to never let things get like that again.

I can even remember what I was wearing that day: a corporate-friendly, midnight-blue-and-white polka-dot dress, with batwing sleeves and a tapered skirt, and black suede high heels. On the outside, I looked the part, but inside I had reached my absolute limit.

My sleep was destroyed – there simply were not enough hours to fit in sleep. Maths may have been my weakest subject at school, but even I knew that if I was going to bed past midnight and waking up before 5am, I certainly was not getting anywhere near the recommended eight hours.

If slowing down was boring before, adding in a high dose of stress and overwhelming anxiety meant that slowing down was no longer even an option. My old job role often saw me trialling fitness devices and sleep tech, and my smartwatch monitoring my sleep was sending me daily warnings.

Things were so bad that if someone had told me to clear my mind and meditate, I would have dismissed them completely. I wasn't just lacking in physical time, I was deficient in mental and emotional time, too. I could not have learnt a new skill, such as meditation, then; I simply did not have the capacity.

If meditation feels impossible to learn, and our monkey brains refuse to stop the chatter, how can we nurture ourselves into more restful states of deep rest, relaxation and even sleep? The fact that you are reading this book means that you already know the answer . . . Sound.

Sound is emotive and healing. It is therapeutic and nourishing. Life-affirming and challenging. It can motivate and spur you forward, streamline your thoughts and cut out the noise. As a meditative tool, sound taps into a part of the process of healing, reflection, nourishment and resetting in a way that words simply can't. It allows people who do not have or want to use their words to delve into a space of self-discovery. Sound is an immersive, full-bodied, somatic experience – a calling into our subconscious, our inner sanctum and our innate wisdom.

WHY YOU NEED THIS BOOK

You need this book because willpower is not enough. You may understand that change is needed and rest is required but do not know exactly how to invite this into your life. Collectively, we are in recovery – from our own past traumas and triggers, the stresses of the pandemic, the changes that the past few years have burdened on us – and willpower alone will not change our mental state. Sometimes a helping hand, a gentle nudge, can make all the difference, and knowing *why* something works may help to cement in our brains that it *is* working.

This is how I see sound – a gentle nudge in the right direction. The warming embrace of change and the non-judgemental holding of space when you need it. Sound also provides the perfect supportive backdrop to help welcome an alternate, replenished way of being. This change does not always have to be drastic. It can be as simple – yet profound – as feeling rested.

Whether this book is a gift to yourself or a gift from someone else, think of these pages as your call to action – your chance to readdress the balance missing from your life. This is a guide to embracing more resonance, a sonic hug to a more peaceful mind and your invitation to rest.

There are many ways to use this book. You do not need to read it in order – you can dip in and out of it and use the exercises as and when you wish. Within its pages, I will show you how to tap into the power of sound, how and why it works, and what you can do at home to elevate your sound bathing experience. You can use these tips every day and combine them with therapeutic sound to bolster the results further. I have also included a glossary and further reading section at the end.

Of course, I could not tell you all about how wonderful sound is without giving you some sound as well. To that end, I have provided a QR code at the back of this book (see page 239) that gives you access to a track that I have recorded with intention. If you have bought this book as a gift to yourself, the audio sound bath included is *my* gift to *you*.

Thank you for letting me take you on this journey to teach you about sound healing. Now let us begin . . .

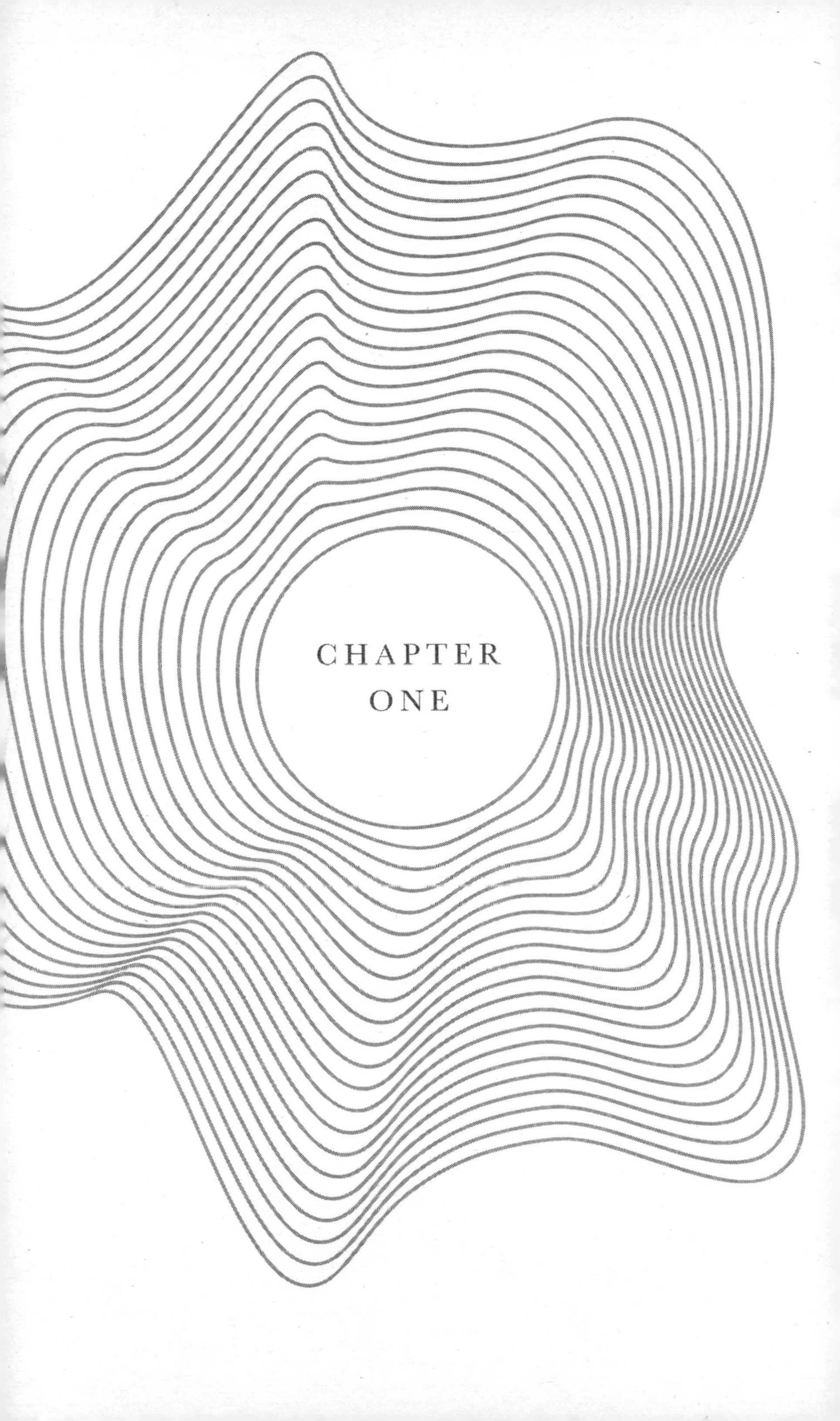

CHAPTER ONE

"He who knows the secret of
sound knows the mystery
of the whole universe."

– Hazrat Inayat Khan

Why Sound?

Before I explore the *why*, I should explain the *what* – what is sound? In simple terms, sound is audible energy, energy that causes molecules to move back and forth, which we call vibrations. These vibrations, if large enough, create waves that your ears can hear. This energy moves through the air but also through liquids and even solid objects. This is why you can hear a baby crying in the next room or a siren outside through your windows and walls.

Everything around you has a vibration – including you. It is just that you can't hear it. Human ears can only pick up a range of 20 to 20,000Hz, and this is only if you have a perfect sense of hearing. Age, damage from listening to loud music, burst ear drums and illness can decrease that range further. Silence is inaudible energy: vibrations that are outside of what your ears can pick up.

Intrinsically we seem to know that we are all made up of vibrations. In the English language, there are plenty of sayings that allude to this inner knowledge. Ever heard the saying "Your vibe attracts your tribe", described how someone "matches your vibe" or even heard someone saying, "We're just really on the same wavelength"?

Chances are that you have heard or used these phrases at least once. Now think for a moment about exactly what they are saying. Of course you know the meaning – but it is funny to think how we use sound terminology to describe situations when we are in sync with each other.

THE EMOTIONAL IMPACT OF SOUND

Sound is highly emotive. This is one of the reasons it works so well and why most of us respond to it in our daily lives without any thought involved. Some sounds calm us while others excite us. A gentle lullaby or the soothing sound of rain can help us fall asleep. Our favourite song can motivate us to strap on our trainers and head off on a run, while a playlist of fast-tempo tracks can even make us train harder in the gym.[1]

Some sounds heighten anxiety. For example, when you are watching a horror film, your senses are engaged before your eyes see anything, the hairs on the back of your neck stand up and you are gripped with a sense of foreboding. Why? Because the suspenseful music tells your body that something is coming, and while you may be sitting in the safety of a cinema or at home on your sofa, your body receives signals that you may need to take action.

Some sounds can be triggering and uncomfortable – ever let your fire alarm battery run out or heard a neighbour's car alarm go off? Within a few short minutes, these sounds can become highly irritating. Sound is so profound that it can even (unfortunately) be used as a form of torture, although this practice is banned by the United Nations Convention Against Torture because sound can evoke such a traumatic response.

This all reiterates that sound is important and heavily intertwined with our sense of wellbeing, whether we consciously recognize it or not.

Sound is also the beginning – the start of creation. The universe started with a "Big Bang", and the rapid hum of lub-dub, lub-dub, lub-dub is what parents eagerly wait to hear at the sonographer's office.

For us, our mother's heartbeat is one of the first sounds we hear, alongside the air whooshing in and out of her lungs and her stomach rumbling. However, the first sound we actually encounter is her pulse, and we *feel* its vibration while our ears are still forming. By the third trimester of pregnancy, we can recognize our mother's voice. Even once we are born and our eyesight is limited, we rely on sound to guide us. We prefer the high-pitched voices of women compared to the lower tones of men.

THE CALMING INFLUENCE OF SOUND

We use sound machines to recreate the familiar sounds of the womb – symbolic of safety – to lull babies back to sleep. The gentle hum also keeps babies asleep and drowns out any noises that could stir and wake them.

White noise has been shown to be useful for adults, too, in terms of concentration. While music can be distracting[2] as your brain listens to the words or gets attached to the melody, white noise provides a variable, random mix of frequencies, like the static of an untuned radio. Playing white noise at a low, steady volume provides your brain with a blanket of background sound that is not distinct enough on its own. This drowns out distractions, enabling your brain's listening centre to focus. Some research has even suggested that white noise could help with memory as well,[3,4] and that, just as with babies, having white noise playing in the background can help adults fall asleep and stay asleep.[5]

White noise is not the only type of relaxing noise out there; there is also pink noise and brown noise. While white noise is a collection of sounds from every frequency, all played at the same time at the same intensity, pink noise is made up of frequencies at different volumes. All of the frequencies are there in pink noise but the lower frequencies are louder. Since the higher pitches are quieter than those in white noise, pink noise is considered to be more relaxing, and studies show that it can induce better-quality slumber.[6]

My personal favourite hued sound is brown noise. Like pink noise, brown noise also favours the lower, deeper frequencies but has an even lower, deeper sound and possesses an almost rumbling quality. Brown noise is sometimes also known as red noise in a nod to red light, which has the lowest frequency on the light spectrum. I find brown noise gloriously peaceful. While there is currently very little data or scientific research into brown noise, people in the neurodiverse community – such as those with ADHD – self-report that they prefer brown noise over white or pink and find it very calming as well.

All of these are examples of why sound can be and is intrinsically calming and beneficial to us from the moment we are born. If you want to learn more about the different coloured sounds, turn to Chapter 4 (page 143).

Clients who may not have consciously used coloured sounds fully understand how soothing a gentle heartbeat (an example of pink noise) can be when it is played on a drum. This particular sound can have the power to settle them into a more peaceful state. Within a few weeks – especially if they come for regular sound baths – they immediately fall into a state of relaxation within seconds of the striker hitting the drum. Their brain has learnt what is about to happen over a course of a few weeks and recognizes it as a cue to relax.

SELF-MEDICATING WITH SOUND

We use sound and music in many ways in our everyday lives to alter or reflect our moods.

When I first started on my sound journey, I remember constantly being met with cynicism. Sceptics from years past could not believe that sound can change your emotions: "What do you mean sound can impact your mood?" Sometimes this question was phrased with curiosity, at other times with sheer disbelief, and in rare instances with distain.

However, when I explained how we all already use music and sounds to self-soothe, and gave examples, each person who posed the question then understood. It was like the metaphorical penny was dropping each time.

- Have you ever put on music to get ready for a night out?
- Do you have a favourite running/gym track?
- Is there a certain song that always makes you teary?
- Has a favourite album helped you through a break-up and made you feel empowered?

If you answered "yes" to any of these questions, you have already used sound or specifically music in these cases to self-medicate. Your favourite gym track helps to motivate you. Your "getting ready" music improves your mood and makes you feel excited. Some music can make you cry and release emotions that you may not have been able to release otherwise. Perhaps you listen to certain songs to give you a sense of strength.

In all these examples, what you have been listening to has changed your mood, allowed for release or motivated you. It increased your heart rate and pumped you up for an activity or occasion. Your reaction to it has not just been in your mind – you felt it in your body, too. You used sound and music in a therapeutic way without realizing it. The terminology or conscious thought was not there but the action was.

As well as connecting us back to ourselves, sound also tethers us to the present. It blocks out the noise of the past and the fears of the future, cuts through worries and tension and brings us back into a moment. Our ancestors possessed this knowledge as well. They knew that sound could impact and change moods and put us into powerful states. We have just misplaced this connection in more recent times.

THE HISTORY OF SOUND HEALING

The funny thing about the modern world is how we keep going back to ancient wisdom. There was a time when those who walked barefoot were seen as uncivilized, but now we call it grounding. Yoga and Ayurveda may have been mocked previously, but now it seems as though there is a yoga studio on every high street. It is as if the modern-day distractions we have created are causing all the stress, burnout and depression . . .

When we look back at numerous ancient cultures, we find that sound was used time and time again as a healing tool by our global ancestors. So much so, that there is even a whole area of study looking into this, called archaeoacoustics.

Acoustic archaeologists have found that ancient humans not only knew the importance of sound but used "super-acoustics" to alter consciousness in rituals and for ceremonies. Examples of this can be found across the world.

Australia

Forty thousand years ago, the Aboriginal people of Australia, one of the oldest human populations in the world, utilized the power of sound healing. We may call it the didgeridoo now but they knew it as the yidaki. Made from termite-hollowed trunks of eucalyptus trees, this wind instrument and the deep resonant sounds it makes were used by these ancient communities to heal a variety of illnesses and even broken bones.

If you think using sound to heal broken bones seems unrealistic, you may be interested to know that it has taken the Western modern world until 2022 to discover that this is possible. A team of researchers at the Royal Melbourne Institute of Technology in Australia found that sound waves can turn stem cells into bone cells.[7]

There is something magical about knowing that the Aboriginal people used sound in this way. The term "Aboriginal" is derived from the Latin *ab origine*, meaning "from the beginning", and I believe that, fundamentally, humans knew the importance of sound, music and rhythm from the beginning.

Some scholars say that this relationship is inevitable because we are – in essence – rhythm. The human body – heartbeats, breath, menstruation, eating, walking – are all rhythmic in nature. The human environment – nature, be that day and night or the changing of the seasons – is also rhythmic. Therefore, life and the very essence of the human experience is about rhythm.

When we are well and happy, we are in harmony; when that harmony is disrupted, we are left with dis-ease. In fact, the basis of esoteric medicine is about how harmony, not function (as we see in Western medicine), is the basis of good health.

Ancient Egypt

In ancient Egypt, the power belonged to the spoken sound. Vowel sounds were seen as so sacred that they were not allowed to be written down in hieroglyphs, and their vocal use was reserved only for priests and those with high-ranking spiritual power or status.

Using a method called toning, Egyptian priests (who doubled up as doctors) used vowel sounds to induce therapeutic benefits. Toning, a technique that is still used today by sound therapists trained in voice work, is when you use your voice to manipulate and extend sounds on the exhale of a breath. The Egyptians believed that the sounds they could generate with toning possessed such potent healing abilities that they made sure their temples, buildings and monuments could amplify these sounds.

Contrary to popular belief, not all pyramids were used as burial sites. Containing acoustic chambers, some were intended as places of sonic healing for the highly privileged living Egyptians at that time instead.

In recent years, some Egyptologists believe that the pyramids themselves, which were built around 2500 BC, were constructed using the power of sound. Sonic drilling, which uses high-frequency sound waves, would be strong enough to cut and shape the great stones that make these vast structures. In fact, this is a method still found today on modern construction sites.

Tuning forks – two-pronged steel bars that have a handle tip at their base and vibrate at length when struck – could also have been used as cutting rods. The reverence and importance of the tuning fork may have been under our noses the whole time.

Hieroglyphics often depicted pharaohs and deities holding a was-sceptre as a symbol of power and status. The sceptre – a straight staff with a forked end – looks suspiciously like . . . a tuning fork. Some historians believe that the was-sceptre may have become a symbol of power due to its ability to cut through hard stones such as granite.

Once the ancient Egyptians cut these gigantic rocks, how did they move them? One possible answer is acoustic levitation, where you lift something into the air using sound waves.

Asia

Moving east of Egypt, other cultures have also embraced the power of their vocals. Chanting is seen across Chinese, Tibetan and Indian traditions. There you have mantras – when you repeatedly chant a word or sound to help you meditate. This is a cornerstone of both Hindu and Buddhist practices. But the main instrument that comes from the region and that we still use today is the mighty gong.

Gongs have been around for the past 4,000 years. They are thought to have originated somewhere between what is now modern-day China and Myanmar, although some suggest they could be even older and that they were originally from Mesopotamia, which is (mainly) modern-day Iraq.

In written history, gongs started to make an appearance in sixth-century China. To have one was considered to be a sign of familial wealth and status and touching one was believed to bring you good fortune. Gongs could also be used in ceremonies and rituals to summon ghosts and spirits and were allegedly used to announce the end of wars.

Ancient Greece

The ancient Greeks also knew a thing or two about organized sound – i.e. music and its importance. Plato believed that the harmony of music was reflective of wider society and could even dictate and change a person's inner character. His view was that "music is a moral law" and could alter the disposition of the listener depending on which instrument was being played and how. He also believed that music was the essence of everything and the source of creativity, saying that it gives "soul to the universe, wings to the mind [and] flight to the imagination".

Aristotle, fellow Greek philosopher and Plato's student, shared similar views. He mused in his writings that flute music could evoke strong emotions. They were not the only Greeks to recognize sound and music in this way. Pythagoras – the same one of triangle fame from your school's maths class – may be one of the best-known mathematicians to have ever lived in the Western world, but he is also credited as being the Father of Music. So much more than entertainment, he believed that music could heal, too, and that certain sounds together could bring the soul into a state of harmony, which could purify the mind and help the physical body maintain good health.

Using his in-depth knowledge of both maths and music, it was Pythagoras who discovered musical intervals. He deduced that certain harmonic ratios produced universally pleasing sounds. Other ratios, he found, were universally displeasing. He prescribed these sonic intervals like medicine and called this musical manipulation "soul-adjustments". Musical intervals are still used today in sound healing. In fact, I use them when I am playing my frosted crystal singing bowls.

Pythagoras also created the concept of "The Music of the Spheres", which is a belief that the planets and the universe all have their own unique signature sound. He concluded that since all of us and everything around us – trees, flowers, animals, and so on – is made of vibrations that can make a sound, there is no reason why larger objects such as the planets would not also have their own individual sounds.

Sound does not travel through space because there is no matter for the vibrations to move though, but plasma waves – created by the energy around the planets – can now be recorded and converted into sound waves. So it turns out that Pythagoras was right to theorize this and the planets do indeed have their own unique sound, or voice, just like us.

England

Closer to home, you have Stonehenge, which was built about 5,000 years ago, with the distinctive circle structure we all recognize built 500 years later in the late Neolithic period. This famous collection of megaliths in Wiltshire, England may be one of the most famous prehistoric monuments in the world, but it was only relatively recently that a team of acoustic archaeologists from the University of Salford concluded that Stonehenge was built with sound in mind and is not just a burial site.[8]

It turns out that the shape of the stones at Stonehenge not only amplify sound but also create a sonic cocoon to encourage attendees into an altered, trance-like state (see page 94). So, if a person was chanting or singing or playing an instrument, although the sounds could be heard clearly, it would seem as though they were coming from all around the listener. Think of this as the Flintstones' version of surround-sound speakers.

Other Countries

In Peru, a pre-Incan civilization built the Chavín de Huántar temple, a series of tunnels and chambers that perfectly encouraged the sonic diffraction (i.e. distorted the sounds) of the Pututus conch (a musical horn). This meant that the conch would sound almost eerie and other-worldly when blown by the priest in prayer and ceremony. Another chamber mimicked the sounds of thunder when large quantities of water were poured into it. This type of sound manipulation shows that our ancestors across the globe truly understood the magic of sound.

Sound manipulation can still be seen at the El Castillo, also known as the Temple of Kukulcán at Chichen Itza in Mexico. This pyramidical structure may have been built by the Mayans over 1,000 years ago, but visitors today can still take part in the clever sound feature that the Mayans built into the temple. Simply clapping your hands once while facing the exterior steps will echo back the chirp of the region's teal-hued, native quetzal bird.

In Malta, you have the Oracle Chamber within the underground Hypogeum of Hal Saflieni temple. The chamber was carved to extend the echo of human voices for up to eight seconds. Using modern technology, one study[9] found that the chamber was also tuned to the resonant double frequency of 70Hz and 114Hz. This was a very significant finding since we now also know that our brains strongly respond to these lower frequencies. In fact, frequencies around 110/111Hz are said to possess a multitude of benefits – including everything from wound healing to boosting parts of the brain where emotional processing happens. It is incredible that a cave built nearly 5,000 years ago could do this.

While most of us have become disconnected from this ancient knowledge, healing with sound is once again getting its moment in the wellbeing spotlight.

THE SOUND RESURGENCE

There are many reasons that sound healing is having a long-awaited renaissance. Timing is definitely one of them. According to the Mental Health Foundation, a staggering three in four UK adults are so stressed out that they have felt unable to cope.[10] Worryingly, one in three UK adults say they have had suicidal feelings due to stress. And, shockingly, 16 in every 100 UK adults have felt so stressed that they have self-harmed. In all categories, it has been worse for women and those aged 18–24, and not only are we deep in the depths of a stress epidemic, but it is getting worse.

Young adults are increasingly finding it hard to calm their nervous systems and build resilience to cope with difficult situations. There is either a skills deficit (unlikely) or modern living is so against our natural biology (more likely) that we are increasingly put in the position of being unable to cope with all the pressures that life puts on us.

Stress is also famous for eating away at sleep. The more stressed we are, the more likely sleeping issues such as insomnia will crop up. A lack of sleep quality and length, in turn, amplifies how stressed we feel. It makes us irritable and cranky, reduces our tolerance, increases our likelihood of being snappy and traps us in a constant loop of chronic stress.

Things are no better across the pond in the USA. Chronic stress – being in a prolonged state of stress – is so prevalent in the United States that one study all but named it as the country's biggest silent killer. Stress wreaks so much damage on both the physical and mental body that the study[11] found that it can lead to sudden cardiac death. Meanwhile, research from Stanford Business School[12] concluded that work-related stress helps to contribute to the death of 120,000 Americans each year. This means that we are all frazzled and looking for a new and easy way to tune out our worries.

Another major factor for sound healing enjoying a rise in popularity is its increased visibility. When the world came to a screeching halt back in March 2020, everyone's way of life was disrupted. Our collective anxiety increased with each moment of fear and uncertainty. Our regular forms of self-soothing and seeking comfort also disappeared; be it the spa, the gym or even the cinema, our go-to methods and outlets for reducing stress were no longer available.

On the flip side, sound workers, myself included, found that we could no longer see clients. There was no possibility for us to get our work out to everyone in the ways we usually would. And then something changed.

I vividly remember receiving an email from magazine site WhoWhatWear.co.uk in April 2020, just one month after the madness started. They asked if I would do an online sound bath for them, thinking it could be a helpful offering for their stressed out, highly worried audience. I immediately said yes, without any idea about how this could possibly take place. Having never worked online, it was a steep and somewhat confusing learning curve.

Four thousand people tuned in and countless messages and requests for more sessions followed. All of a sudden it did not matter that I lived in London and someone who wanted to enjoy the benefits of sound lived in Manchester, Belfast or even Dubai, as everyone could join in. Zoom, IGTV and social media meant that people did not need to have a wellness studio that offered sound baths near them. They could now try a sound bath – for free – at home. Suddenly people who had never heard of sound healing were able to experience it for the first time. This was a key moment for increasing the visibility of sound as a healing modality.

We were not the only ones to take an interest in sound meditation. As celebrity interest from individuals such as Madonna and Kendall Jenner has grown, this has further exposed sound healing to more people. Of course, celebrity interest does not immediately translate into everyone else wanting to try something, and increased visibility can only take you so far.

One of the things I love most about sound work is how everyone can take part. It can easily become a feature of anyone's self-care practice. Powerful but essentially passive, sound meditation requires no effort on the part of the listener, attendee or client. You do not need to master a special skill to enjoy and benefit from sound work, and there is no learning or effort involved other than turning up or tuning in. This is even more vital when you are in a state of overwhelm and stress.

Your practitioner or therapist does all the hard work for you. The learning and acquisition of skills is our responsibility, and it is we who need to master the correct playing techniques that will change your brainwaves (see page 89). In a session or sound bath, you simply lie down and listen.

Another unique selling point of sound is that it works – and can be felt and experienced – immediately. You do not need to put hundreds of hours into it to benefit. Multiple sessions across multiple weeks will boost the benefits, of course, but just one session can make a difference. It is not uncommon for clients to report feeling a sense of calm, rest and relaxation immediately after their first experience. Just one sound bath can prove to someone trying sound healing for the first time that it is not just an abstract idea.

In an increasingly busy world where we are always connected or having to do *more*, there is a sense of escapism when we do not need to do anything at all. This is like going for a massage – your body benefits greatly and all you did was turn up.

THE PRIVACY OF SOUND

Another huge draw for sound, especially sound therapy, is the complete privacy it affords. While every type of therapist across all modalities will keep their discussion with their client completely confidential, for some, simply revealing what they are struggling with can be overwhelming. At other times the client may not have the words or wish to say the words out loud. With sound work, you do not have to verbalize the problem to address it.

In a sound therapy session, you can explore how you are feeling in an abstract way. This allows you to keep what is happening or troubling you to yourself. There is no need to share anything you do not wish to share, but that will not stop you from enjoying and receiving the maximum benefits from the session.

Feeling indecisive about an important business decision that you can't talk about? Perhaps you have not quite pinpointed what is making you feel anxious, or you simply do not want to talk about something but just want the tension it is causing you to go away? If so, then therapeutic sound can help.

Sessions are designed to enable you to access your inner mind, the deeper levels of your subconscious, so the exact words of what you are looking to shift or uncover are not needed. This is probably why I often see clients express how shocked they are by what comes up for them in sessions. Freed from using words, we also tend to be more honest with ourselves about what we need to work on.

There is no judgement from a practitioner, of course, but sometimes the restrictions and judgements are present, and these are the ones a client brings to their sessions themselves. Not having to speak directly about an issue they want to address can therefore lead to deeper levels of insight.

Avoiding articulating something you wish to work on is also helpful when processing grief. For many of us, losing a loved one or experiencing any type of loss can be overwhelming. Sound has the ability to hold us in an aural blanket of safety and comfort to help us on the journey to recovery without expressing words.

Perhaps this is why so many of the ancient sites we explored in these pages already had sound intertwined with places of burial, ceremony and ritual.

Another reason for the renewed interest in sound healing at the moment is the collective cultural shift toward being more open to holistic health remedies. Why pop a pill if a few small lifestyle changes could afford you a better quality of life? A deeper awareness and respect for the importance of mental health and this ancient tool for healing is allowing a modern audience to recognize they do not simply have to "put up" with feeling stressed or anxious. These energy-draining states of being do not have to be forever; you can change them. You are allowed to feel rested, relaxed and happy.

In recent years there has also been a lot of talk about manifestation and how changing your thought processes can lead to the accomplishment of your goals. But while manifestation and the accompanying visualizations of a new life and reality seem to come easily to some, others have struggled to get the results they want. However, a new understanding in those communities has started to emerge. People are beginning to recognize that if your manifestations have not yet worked, it is not usually due to a lack of effort. Far from it.

Instead, there is a growing understanding among manifesters that perhaps the real reason that things have not, and are not, working out for them in terms of their goals could be because of an underlying disconnect between their mind and their body. Maybe their New Year's resolutions, or a new wellness plan or goal they are trying to achieve, has not come to fruition because their body – and therefore their mind – is stuck in fight-or-flight mode.

Long-term stress and anxiety can cloud your thoughts, as we will go on to discuss in the coming pages. If you are chronically in survival mode then you cannot create from this place. And manifestation is all about creating a new reality.

These individuals seek out sound healing to get them over the finish line with their life ambitions.

Perhaps it is the same for you – maybe you have picked up this book because your goals are not progressing. Luckily, in this book you will learn how sound supports self-inquiry and discovery, and how changing the state your body is in will free the mind so that you can bring your manifestations to life.

You will come to understand how sound healing connects you to your inner wisdom and vision, and reawakens your true desires. You will realize all the ways sound allows for your nervous system to feel safe. From there it will be easier to make life changes and manifest your objectives. But first, you have to start from a point of emotional and mental regulation.

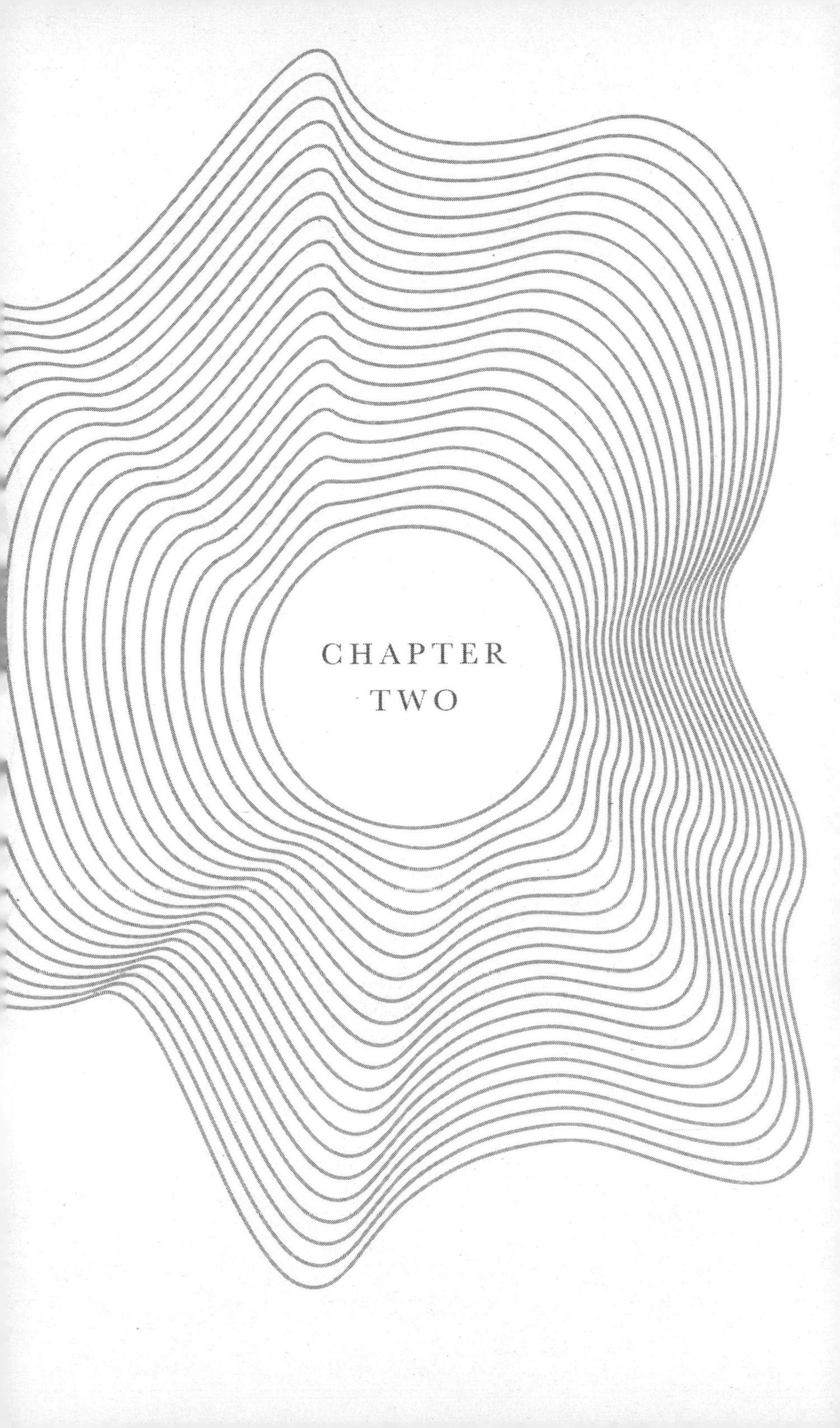

CHAPTER TWO

"If you want to find the secrets of the universe, think in terms of energy, frequency and vibration."

– Nikola Tesla

What is Sound Healing?

Healing – "the process of becoming or making something or somebody healthy again. The act of getting better." This is how the dictionary defines healing. That it is recovery, and insinuates something has happened that made you unwell. You will notice there is no clarification of whether this process is regarding something mental, emotional or physical; by not doing so, the word "healing" encompasses them all.

Sound healing is rest, but it can also be recovery. It is a full-bodied, somatic experience that enables you to return to health when there has been emotional and mental pain, suffering or discomfort. Sound healing can also help with physical pain, but we will get to that later. For now, I want to concentrate on the pain you can't see, especially since sound healing enables you to look inward.

In the Western world, there is a disconnect in the way we view healing. We have fallen into the trap where we are led to believe that we have to constantly seek the answers outside of ourselves. That everything must be intellectualized in order for it to be understood – solved. I am a huge fan of psychotherapy, and talking about your struggles is essential at times, but it is also important to allow your body to process what it has gone through or is indeed still going through.

Using the power of sound waves, sound healing enables you to get to the processing stage. It is a holistic treatment that signs the permission slip for rest and recovery, no matter how you have experienced your harm. It holds you in a cocoon of sonic waves so you can let go of what you no longer need.

For many of us, healing only seems to get the green light when we are suffering or have suffered something physical. Break a bone or twist an ankle, and rest and recovery are expected. Go through a break-up, experience stress at work or react to one of life's challenges with anxiety, and there is an expectancy that you should "keep your chin up", "throw yourself into something new" or "keep yourself busy", even if the mental and emotional hurt triggers a subsequent physical reaction as well.

"Others have it worse", "Don't dwell on the past" or "Look on the bright side" may also be phrases you have heard when the pain is mental or emotional. This completely ignores and fails to recognize the fact that your nervous system is hugely out of whack – and you can't simply think your way out of it. Nor can you talk yourself into a rest and digest, parasympathetic state. The parasympathetic state is the opposite of the fight-or-flight response and tells you when you are safe – it takes the cues from the world around you to signal safety. You can't think your way to it; you have to *feel* your way into it.

Since we are talking about healing, I should add that this does not mean that I am a healer. I am not. I am a facilitator. A teacher. I am there to make space for you to reach your own healing; I am there to guide you to the answers that are already within you.

THE MIND–BODY LINK

Your body and your mind are intrinsically linked. If you think of your brain and your body as two destinations connected by a road, you must remember that the traffic flows in *both* directions, and this flow is highly important.

When your mind is excited, you feel it in your body through butterflies in your stomach. When you are angry, your body feels hot. The same also applies in reverse. When you step outside of the chilly shadows on a spring day and feel the sun's warmth kissing your skin, you can't help but smile and feel joy in that moment.

Body language can even directly impact how you are feeling. Softening the face and smiling, lifting our arms up above our heads in celebratory pose or even sitting up straighter has been scientifically shown to lift our mood.[13 14]

If you are in a constant state of stress and anxiety, simply thinking a different thought will not change your reality, because memories are not just stored in your brain but also in your body. To change the way you are thinking, you have to change the way you are feeling – you can't just talk or think your way out of stress or trauma. First your body must know it is safe; you have to give it permission to come out of fight-or-flight.

In recent years, neuroscientists looking into anxiety have started to confirm what sound workers have known all along. That anxiety and chronic stress are feeling states that then lead to overthinking and worrying: body first, mind second. I am not saying that it does not work the other way, too. But long after your mind has perceived a threat, the remnants could remain inside you – sending signals *back* to your brain saying that the danger has not yet passed. This encourages you to stay on high alert, technically known as hyperarousal.

Another incredible discovery has been that the heart also contains cells that store memory independently of the brain. This helps to explain why break-ups and grief are felt across the body. And why, when faced with these life-defining or indeed life-shattering moments, we can't just think our way back to feelings of happiness.

Maybe this is why I hate the following phrase with a passion: "What doesn't kill you makes you stronger." In my opinion, this is one of the most toxic statements to exist in the English language. It is designed to make us brush over our stresses, triggers and boundaries and to keep going. To keep being productive; to slap on a happy mask and reply, "Yeah, I'm fine," when you are not.

This phrase is completely false, too – a lie in terms of your body – because what doesn't kill you wrecks your nervous system. It strips you of your ability to remain calm and in control of your feelings, changes your default state and sits you in a place of anxiety, anger and fear. It dysregulates you.

Think about it like this: are you more prone to misunderstanding the tone of a harmless text from a colleague, boss or even friend when you are already angry or annoyed? In the same way as you are more likely to snap at a partner when you are in a heightened state of stress, even if all they have done is ask you a simple question such as, "How was your day?"

You can become so stressed out and anxious that it becomes your default, and when that happens you may feel like you have lost control of yourself. Your body stays in a constant state of alert, so you lose control of your day to day.

A simple request at work becomes an overwhelming, stress-fuelled event, or choosing what to cook and eat for dinner becomes a set of insurmountable decisions. You start to lose the sense of connection with yourself and inevitably, as a result, you lose your connection to others. So how do you fix it? How do you heal this connection?

Rebalancing Your Body–Mind Connection

I think, in the world we live in, the most crucial first step is continuously being missed out. That is, before any attempt at healing, you need to first regulate your nervous system and bring your body back to neutral. You have to build the foundations again, or maybe lay them down for the first time. Otherwise you are just decorating – applying new wallpaper on patched-up walls. Things may seem good initially, but it just takes one storm, one big gust of wind, for the entire wall to crumble – fancy wallpaper included.

Regulating your nervous system ensures you have a strong base to begin with. Then, building a strong vagal tone (more on this on page 102) ensures that each brick has cement and not sand in between.

If you miss out this step, "It's not so bad", "I can manage this" and "Things will get better" will all be statements that you never believe. You simply will not have the ability to accept them and your body will be at loggerheads with your brain. No matter how often your brain says, "It is okay," your body will continue to send messages to override this, keeping you in fight, flight, freeze or fawn (see pages 40–1). It will tell you, "Things *are* bad," "You *cannot* manage" and "Things will *never* get better."

The fact is, the stories you tell yourself about what is going on in a situation and how you interpret an event, how you see yourself, how you perceive the actions of others, and how you think they will perceive your actions, are directly impacted by what your nervous system is doing. And this is where sound comes in.

Sound is an excellent vessel for meeting you where you are. It does not matter how dysregulated your nervous system is, how anxious or stressed you feel, how wired you are when you walk into a sound bath. Sound will hold you where you are and provide space for you to rest into it, and this will slowly enable you to let go of anything that is no longer serving you. Sound allows the opportunity for you to heal yourself.

BALANCING YOUR EMOTIONS

This is a good time for a short pause so that I can address emotions. We can be quick to vilify what we perceive as "negative". That is not what I want to do here, nor do I want us to fall into the saccharine trap of toxic positivity.

You are not supposed to live a life without any challenging emotions. You can't be happy and stress-free all the time, and that is okay. Life is about ups and downs. There will always be highs and lows. Once again, this is okay.

It is important to remember that your thoughts and feelings of anxiety, fear and stress serve a purpose. Some stress is good for you, as, indeed, is anger and anxiety. They are all warning signs and indicate that you need to act. They tell you something about where you are. That your current circumstance or situation is wrong, that you may be unsafe – that a personal boundary has been crossed and something needs to change.

In essence, anger is a messenger. It is an emotion designed to protect you and keep you safe. It is a response to something that your mind and body perceives as unfair or threatening. Anger can also feel like a safer alternative emotion to express than, say, pain or sadness.

The hormones that flood your system when you are feeling like this serve a purpose, too. They enable your body to react and are designed to make you faster in emergencies. It is the whole sabre-tooth tiger chasing our ancestors scenario. This is why some stress is needed. Anxiety is not to be feared either. The discomfort shows you that change is required. These signals are essential for your wellbeing. Short-term stress demands action and then it leaves, and your body recovers and goes back to a healthy baseline.

What we are addressing in this book is prolonged stress – chronic, or long-term, acute stress that has taken over your entire life. This type of stress produces negative thought processes that consume you and stop you from living the life you deserve – it controls your body and therefore your mind. Anxiety that has interwoven itself into the fabric of your daily activities and interactions and is now ruling you. This is what we are here to change – and heal.

YOUR STRESS RESPONSE

When you are stressed, your body responds by activating your sympathetic nervous system. This is more commonly known as the fight or flight response. These days, it is recognized that there are actually four different responses that can be activated in times of stress: fight, flight, freeze and fawn.

The first two states (fight and flight) elicit a very physical active reaction. The third (freeze) leads to a more passive physical reaction, and the final one (fawn) results in a more physiological response. What these have in common is that they all deeply impact and influence the way you think and see the world. Being stuck in any of these states for a long time will cloud your judgement and muddy your daily thoughts. You will see things as *you* are, not as *they* are.

Fight

Fight energy will make you view everything with a negative stance. Misunderstandings will happen and you will lose the ability to see things as they truly are. An example of this is when your boss asks to have a quick word and you immediately start to panic. The narrative in your head is that it is going to be about something "bad", that you must have "done something wrong", that they will blame you for something.

Your nervous system is deep in survival mode, so you perceive even a neutral experience to be a threat or something that is unsafe. When you are in this frame of mind, your memory can be affected as well – you may find yourself in a situation where recollections may vary, when you remember things with a much harsher or negative viewpoint.

Flight

Flight energy will always have you in a state of running, always on the go. Looking back at my old life, this is exactly where I was most of the time. Here, you pack more in, you over-commit to try to use up that energy, you try to do more and more to overcome the stress you are feeling. This leads to statements such as "Once I've done this . . .", "I just need to get past this and this . . ." and "Things will get better once I've completed . . ."

Freeze

The Freeze stress response usually kicks in when your body thinks it can't fight or flee your way out of a situation. It can be a combination of both a sympathetic response (fast heartbeat) and the start of the Dorsal Vagal state (holding your breath, completely immobile). Here you feel like you have no choices left and that nothing can be done. A sense of hopelessness may start to crop up and, over time, stick.

Fawn

Like Freeze, the Fawn response appears when your body does not think it can fight or flight its way out of the situation. Your stress becomes so acute that you fall into conflict-avoidance mode – also known as "extreme people pleasing". Fawn energy will trigger you to avoid and suppress your needs. You will try to "soldier on" and you will put yourself at the bottom of your priority list.

HEALING YOUR MIND

Your thoughts are influenced by your nervous system. When you are in one of these stressed states, you are robbed of the ability to see things clearly. You can't see the bigger picture. Your nervous system taints your view of the world so what you do see is clouded. When you are triggered, you also can't be creative or innovative and your problem-solving capabilities become out of reach.

Regulating the nervous system gets you out of these heightened, alert states, and in doing so, it allows you to have a clearer head. You return to a more in-flow state, which brings with it out-of-the-box thinking.

It is no surprise that so many of my clients do not just experience an absence of stress from their lives when they build up a regular sound healing practice, but they also find that their lives become more fulfilling. Suddenly, they have the mental space to incorporate other positive changes as well. This is why, straight after a sound bath, you may be flooded with ideas. My own business plan came to me after my first-ever drum circle many years back. I "awoke" from it with a vision as clear as a blue-sky day. Grabbing my notepad, my ideas poured onto the paper via the ink.

Being able to think rationally is also the reason that, after a sound bath, things that seemed insurmountable suddenly start to appear manageable. Obstacles that blocked you from feeling a sense of peace get a bit smaller and your view becomes a little clearer. Eventually you will find these obstacles can even disappear completely and you begin to get answers to problems that could not be solved before.

You regain the ability to see things as they really are. A calm, well-regulated nervous system empowers you and teaches you that you can keep yourself safe. Things then begin to get easier.

WHAT IS A SOUND BATH?

Okay, first things first – just because there is the word *bath* in it does not mean there is any water involved. Unlike a relaxing bubble-filled soak, a sound bath is a holistic treatment that uses sound, frequency and vibrations to take you to a deeply restful, physical and mental space. Sound bath is also the blanket umbrella term given to any type of sound healing or sound therapy session.

A sound bath differs from music therapy in that only sounds and notes are used, not melodies, beats or lyrics. Instead, therapeutic-grade instruments are played in a way that slow down your brainwaves, indicating to your body that you are safe. This in turn activates your parasympathetic nervous system. The result is that you start breathing more slowly and your heart rate slows down. Your focus goes inward. You start to detach from the physical world around you and your mind enters a deeper state of consciousness. This allows you to reach a deep state of meditation. It is an invitation to rest – a way to heal and to put down and recover from the stress you may be carrying around with you.

But *why* is it called a sound bath? Because the sound waves from the instruments are said to bathe you in their meditative vibrations; they wash over you, enveloping you, healing you. That said, it can be beneficial to think of a sound meditation like a regular bath. Getting into an actual water-filled bath at home requires setting time aside. Unlike a shower, a bath can feel more like an experience. You may light some scented candles, pick up a book and fill the water with body salts or pour in a skin-softening bath oil. A bath allows you to be present in the moment and enjoy just being.

In many ways, a sound bath is the same. You are required to make time for yourself and have permission to press pause on everything else happening in your life. It is an invitation and a gift to just be. Nothing productive in the typical sense of the word is required from you. A sound bath affords you the time to be a human *being* rather than a human *doing*.

WHAT HAPPENS IN A SOUND HEALING SESSION?

Typically, a sound bath will involve you lying in a comfortable position, although you can also enjoy sound while you are seated. The main thing is that you are in a position that feels safe and pleasant. In Chapter 6, we cover some of my favourite ways to lie down in a session, but most classes will take place on long cushions with pillows and blankets.

This type of deep meditative work cools down your body temperature, so blankets are a must as you will likely start to feel cold. Because you are lying down for an extended period, I always recommend putting a long bolster (or any cushions you have) underneath your knees to help support your back more effectively. If you are blessed with long legs, also pop a cushion under your feet to elevate them. This feels especially good if you have been on your feet all day.

Individual sessions, especially if these are sound therapy sessions, tend to be on a treatment bed. The elevated set-up makes it easier for the practitioner to get around you and treat on (or very near) your body when needed.

INSTRUMENTS OF SOUND HEALING

Speak to any sound therapist or practitioner and you will find that instruments are our weakness, our kryptonite – we love them and would have them all if space was not an issue: one of each and in every colour and note. I often have different instruments with me on different days and may change the instruments I use from one class to the next on the same day. This is why no two sound healing sessions are ever the same.

The main instruments used for sound baths tend to be crystal singing bowls (the large, white frosted ones as well as the colourful alchemy ones), Himalayan singing bowls, drums and gongs. All these instruments can help you achieve what sound therapists called an Altered State of Consciousness (ASC – which we will explore in more detail on page 94) and help your nervous system downregulate, i.e. calm you down.

The type of session you go to is completely up to you. Some sound baths will feature several instruments so you may not have the option to pick, but others may focus on only one particular instrument. If the options are available, try out as many different instruments as you can. Some private clients have immediately fallen in love with the Himalayan bowls, while others love the drum. Some can't imagine attending a session that did not have crystal singing bowls. Allow yourself to explore so that you can find what works best for you.

For years I thought I hated the sound of the gong, finding it aggressive, confrontational and masculine, which in itself is telling. The instruments we do not like, or the sessions we find uncomfortable, can often reveal something fundamental about ourselves.

The reason I hated the gong at that time, I now understand, is because I was very much in my masculine energy – I was out of balance. The powerful humming of the gong was pushing me further into the extreme of where I was at that time. I disliked the gong so much that I did not include it in my initial sound practitioner training. Later, once I was more balanced, I found myself drawn to it. Now, of course, I love it.

Do not be shy to try out different teachers either, as every practitioner will bring their own innate wisdom, life experience and energy into their sessions. I am drawn to certain styles of play, too, when I attend a sound bath. By exploring different teachers and a range of instruments you can find which works best for you.

A gentle reminder here, though. During your sound bath, do not try to work out what made *that* noise or try to take a peek when you hear something that sounds strange, different or even familiar. This is your conscious brain bringing you back into thinking mode, into categorizing and understanding. You are in a sound bath to switch off, to do the exact opposite, so let yourself go with whichever sounds you hear, and feel them, no matter what instrument is making them. Afterwards, you will be able to ask your practitioner. I always let clients know that they are welcome to ask questions after a session, share what they have just experienced or even talk about what they found challenging. But in the middle of a session, give yourself permission to be in the moment.

Himalayan Singing Bowls

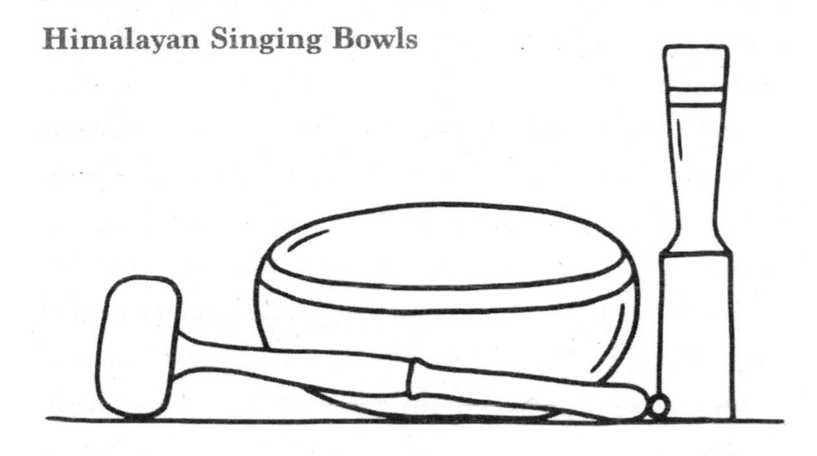

These bronze-hued metal bowls are what we call the sonic cuddle of the sound healing world. Full of harmonics, singing bowls produce warm, rich and comforting sounds to lull you gently into an alpha-brainwave state. I find these bowls are great for someone trying sound for the first time, as there is a richness to their tones that scoops you up and holds you.

The bowls are struck (donged) with long mallets and played along the rim. The vibrations they produce are powerful, so if you have metal fillings you may want to tell the practitioner not to get too close to you. These bowls can also make in-person sound healing unsuitable for anyone with a pacemaker, as the vibrations can interfere with the rhythm of the machinery.

One client who had recently given birth found that the bowls, while a comfortable metre and a half away, were making the breast milk inside her body vibrate. After this, we made sure she expressed before her sessions.

Himalayan bowls can also be placed directly on the body for what we call vibroacoustic massage, helping to release tense muscles. I have seen many gimmicky videos of people sitting inside giant bowls or donging them on their head. Personally, I would advise against that. The vibrations pass the dermis of the skin and into the body, making them very powerful. This is why I only like to use the bowls on the back of the body and on the limbs – avoiding proximity to the major organs.

Sometimes it can feel as though the practitioner could use a bit more force when using the bowls, but vibroacoustic massage is deceptively strong and the DOMS (delayed onset of muscular soreness) the day after will be testament to that. This is why a mixed session where the treatment is broken into two parts featuring a soundscape as well as a massage can be a good idea, especially if it is your first sound massage treatment. This enables you to see how it feels on your body.

These metal bowls, thought to have started life initially as food dishes in Northern India and Nepal, are sometimes also (incorrectly) referred to as Tibetan bowls. In recent years, many people from the Tibetan community have called out against the bowls being aligned with Tibet, voicing that calling them so is a deeply harmful commodification of their culture. Since calling them Himalayan bowls is not only more accurate, but also avoids cultural harm, I always refer to them in this way.

Himalayan singing bowls are also victim to some creative misinformation. The idea that these bowls are made of seven different metals is a myth that has been extensively debunked by Oxford University, no less, and yet it continues.

The "legend" goes that each bowl is made from a composition of different metals to symbolize different planets. We have gold for the Sun, silver for the Moon, copper for Venus, iron for Mars, tin for Jupiter, and mercury and lead for Saturn. Not only would having bowls containing gold whack up the price, but consistent testing has shown that all hand-hammered bowls are actually made from the metal alloy bronze, which is a mixture of copper and tin. This combination of metals (which is higher in tin) is called bell metal bronze – it produces a beautiful sound and has been used in bell making for thousands of years. The bowls' metal composition also means that technically Himalayan singing bowls are part of the bell and gong family, and are in fact standing bells rather than bowls.

Lower-quality bowls produced with machines are made from brass, which is a combination of copper and zinc. These do not sound as good in my opinion and are more likely to be found in crystal shops and yoga studios than a good instrument outlet.

There are also some myths about how the bowls are made. I do not blame the sellers for their creativity in trying to add some magic to people's purchases, but I am sorry to say that the chances that your bowl was made by the light of the full moon or while monks chanted over it are slim to none.

Of course, the myths will not matter to many of you, since you will not be buying instruments, and nor do I recommend you buy them either. Apart from the expense, playing a bowl or therapeutic instrument puts you in a different headspace to when you are just listening to that instrument being played. This defeats the purpose of why you are listening in the first place. The only exception to this is the drum, and we will get to that soon.

With these myths busted, does this diminish the healing properties of these beautiful singing bowls or inhibit their ability to calm a wired nervous system? No, not at all. Even if some of the romance of storytelling is lost, these bowls continue to be a useful prop in sound healing.

Crystal Singing Bowls

These are not often the first instrument that people think of when they imagine a sound bath. There are two types of crystal singing bowls: white, frosted quartz crystal singing bowls and the colourful alchemy singing bowls. Not only do these look different, but they sing differently, and in my opinion serve slightly different purposes.

Let us start with the colourful alchemy bowls. Made of 99.99 per cent quartz, they are fused with trace amounts of different gems, such as emerald and ruby, and metals such as gold and platinum to give them their famous beautiful hues. They may be lighter and smaller than other types of singing bowls but they are also eye-wateringly expensive. If you get too close to them at a sound bath, their cost is the reason why sound practitioners may get nervous. One accidental touch could easily result in thousands of pounds worth of damage.

Like the Himalayan bowls, these crystal singing bowls can be full of harmonics, meaning they may create several notes rather than one single note. The lowest note (the fundamental note) will be marked on the bowl. These types of bowls have a more piercing tone and are perfect for creating binaural beats in a session. This is where you hear that "wah-wah-wah" sound.

Most crystal singing bowls have a rounded bottom and sit on a rubber ring to keep them in place. Practitioner bowls are also made of crystal but are goblet-shaped with a quartz rod handle, making them perfect for holding and walking around a room.

The frosted quartz bowls tend to be larger, definitely heavier (just ask my biceps), and predominately white in colour. These are the original crystal singing bowls (the alchemy ones came later). They have a rougher exterior texture and inside are smooth and glass-like. Made from 99.997 per cent quartz, these instruments came into existence purely by chance. Originally a waste by-product from Silicon Valley, quartz crystal bowls (which can withstand very high temperatures) acted as crucibles, a melting pot used in the manufacture of computer microchips. The discovery that they were extremely resonant and produced beautiful sounds that could be used for relaxation was accidental and makes them a contemporary sound healing instrument.

These bowls are pure toned, which means they lack the harmonics of the other bowls we have mentioned so far. Each bowl only contains one note. When certain notes are played together, they create incredibly spellbinding and highly emotive musical intervals.

I love using these types of bowls and creating the intervals with them. Frosted crystal bowls have such an expansive sound that they are useful tools for deep introspection, for working through and overcoming an issue and releasing traumas. Their size and pure sound means that they also produce much louder sounds than other types of singing bowls, making them perfect for larger groups.

Drums

Out of everything I have in my kit, my most unique relationship is with my frame drums. Physically speaking, they are the most challenging to play, and mentally they are the most challenging to not fall into a trance with. With such difficulties, it may seem odd for me to keep them in my kit – yet some of the most surreal, euphoric, spiritual, creative, transformative, calming experiences that I, or clients, have had all involve the drum.

The drum needs to be an in-person experience. You need to feel its vibrations and let the energy of the rhythm move the energy within you. Like the Himalayan singing bowls, frame drums can be used for vibrational massage as well. But, unlike the bowls, the drums are not placed on the body, and are instead just played very near. The vibrations produced by them are strong, making the drum an excellent massage tool for combating aches and pains.

Drums are brilliant for bringing focus, self-awareness and mental clarity. They also have the ability to take you into a trance-like state more quickly than other instruments. It is

no surprise really that drums show up in every culture across the planet, often seen as sacred and used for ceremony, war, dance, celebration and for invoking spiritual experiences. The drum is so good at taking someone into an Altered State of Consciousness that it is sometimes hard for me as a practitioner to keep my focus and not be swept up with it.

Gongs

The mighty gong is a powerful, intense, highly emotive instrument that is excellent for moving stuck emotions and thoughts and banishing self-limiting beliefs. Gongs are also known as Tam-Tams. Before entering the world of sound healing you may have seen these struck dramatically before an old film starts. (If you do not know what I am talking about, type in "Rank films intro" into YouTube, where you will find an oiled-up, shirtless man striking a gong as if his life depends on it.) Fortunately, modern sound practitioners are neither required to be shirtless to play the gong, nor do we strike it with a comedic caricature force.

Falling in the percussion family, gongs are disc-shaped and made from metal. They have been used in European orchestras since the 1790s and for thousands of years in the East. Just like the Himalayan bowls (which are considered to be bowled gongs), standing gongs (as they are also known) can be made from either bronze or brass.

Depending on the mallets and strikers used, a gong will produce an array of different tones and frequencies: high tones are designed to stimulate you and low tones to relax you.

Gongs can be categorized in a few different ways.

- **Flat gongs** – these do not have a rim and as the name suggests they are fully flat. These may also be called wind gongs.
- **Nipple gongs** – also known as a bossed gong, these have a raised centre or knob in the middle and tend to contain lower frequencies.
- **Chau gongs** – these are nearly flat, the face of the gong being slightly concave. Centuries ago, these were the types of gongs that would have been used in processions and on important government business. When two high-ranking officials would meet, the gong would be hit, the more strikes indicating who ranked higher.
- **Flanged gongs** – these are the ones that have a wide, curved rim.
- **Symphonic gongs** – these gongs contain a wide variety of rich tones, making them the perfect addition in to orchestras as well as sound healing sessions.
- **Tuned gongs** – these gongs are tuned to the fundamental frequency of a planet, star or celestial body, the calculations of which were based off the work of Swiss mathematician and musicologist Hans Cousto.

Some categories can cross over, too. For example, a Patiste symphonic gong is also an example of a flanged gong because it has a curved rim.

Sound baths that predominantly or only use gongs can also be called gong baths, but a sound bath may not necessarily be a gong bath.

Therapeutic Percussion

This is where chimes, shakers, rattles, thunder tubes, ocean drums and rainsticks come in, as the gentle, rumbling awakening towards the end of the sessions. Not everyone loves this part, but it does serve a purpose. These lower-frequency, gravely sounds, and the way they are played, are designed to bring you back to a more conscious alertness. They lull you back from the dreamy, deep meditative state to leave you feeling more awake. People who do not like therapeutic percussion do not like it for those very reasons, as it signifies the session is coming to an end. Compared to how the instruments that precede them are played, the therapeutic percussion sequence is rousing – and it is meant to be.

For the audio that comes with this book (see page 239 for the QR code), I have kept the percussion gentle and fairly short, so if you wish to fall asleep straight afterwards, you can. The accompanying audio is, of course, not suitable for listening when you are driving or need peak concentration.

The Best of the Rest

Many other types of instruments and sound healing tools also feature in sound baths but there are too many for me to delve into here. You may see monochords (an ancient rectangular stringed instrument), handpans (also known as steel tongue drums), Shruti boxes (Indian drone), tuning forks, harps, crystal pyramids, xylophones and even didgeridoos being played in sessions.

None of these instruments are better or worse than the others. It is all about personal preference. How your brain interprets and receives sound is unique. Your practitioner's main objective is to bring you into a place of calm. You will not resonate with every instrument and that is okay. For me, the sound of a tingsha (a small hand bell where two cymbals are attached by a piece of leather) will never be calming – I find them shrill and agitating. So there may be sounds you do not like and some instruments you love. Just explore with an open mind.

I should also add here that no two sessions are the same, even if your practitioner is using the same instruments. The energy of the attendees and even the energy of your therapist on a particular day, as well as the weather and the temperature of the room, will all impact on how the sound is produced and how you hear it. On the days I have back-to-back classes, I often change which instruments I use and how I use them. This is why it is important not to have a set idea of what a sound bath may be like. Keep yourself open to receiving sound as it comes.

SILENCE IS GOLDEN

As we have already explored, sound is simply audible energy, but there is so much power in the inaudible energy that follows a sound bath that you should not disregard this. I always include a minute or two – sometimes even up to five minutes – of silence in every sound bath I facilitate. Everyone is reminded of the silence that will close the sound bath, whether this is their first session with me or their hundredth, and they are told to stay in it and lean into it.

For the busy-minded, this can initially seem like a waste of time. Why would a lack of sound have any benefit in a *sound* healing experience? Surely the silence does nothing, right? Wrong. The silence that follows a sound bath is loaded with energy and can feel so loud that it becomes deafening. This can be incredibly powerful, and I do not say that lightly.

Whether you are alone in a one-to-one session, doing a sound bath at home or in a group setting, the silence lets you truly be with yourself. And with that, it invites a type of reflection that we do not often honour. The silence that follows a sound bath allows you time to really sit with yourself – to look inward, to be present in what you have just experienced and really take that all in. If you have accessed insight in your session or you are filled with a sense of peace and gratitude, sit with it, let your thoughts wander and really let yourself go into it.

If you have experienced a trance-like ASC state (more on this in Chapter 3), and therefore achieved a theta-dominant brainwave state, your brain will need time to slowly come back up to a beta-dominant waking state. You should not immediately go from being in a deep meditation back to normal chit-chat. The silence provides the bridge between the two. It is also a beautiful contrast to the sound and delivers a sense of balance. Silence is the yin to the yang of sound, the dark to the light. You can't have one without the other; their relationship is one of perfect symbiosis.

I am not surprised when first-time clients comment on how they can't remember when they last sat in silence. Either we are with people or we are connected by our tech to people. At other times when you may be alone, you will find that you fill your space with the din of the TV, the radio and music.

You may even do this to keep a sense of loneliness at bay, and that would be valid. Silence can remind us of a lack of connection, a moment without community. But without some silence in your life, it could also be keeping you from connecting with yourself – stopping you from hearing your inner voice, not allowing you to tune in to your essence, to find yourself.

Sound Bite

Less is More

A few words of advice for people who attend sound baths and those who facilitate them. In a world where social media is filled with images of sound baths happening with hundreds of bowls and an army of gongs, it can feel like you need an orchestra to get the maximum benefits. This is simply not true.

Attendees, trust me when I say that I could entrain you with just one bowl, playing one note. I know this because I have done it. So do not be tricked into thinking that your practitioner needs hundreds of instruments in their sonic kit for you to get the most out of your session.

Practitioners, if art is how we decorate space, and music (or, rather, sound in this case) is how we decorate time, remember that no artist uses every single colour in their palette for each painting. The same applies here. You do not need to use the flute, the drum, every chime, gong and bowl you have in your collection in every session. Less can be more. Using the right notes and frequencies can be more powerful than squeezing in as many different sounds as possible. If you imagine a sound bath is like an audible rainbow, then using too many colours from your sonic palette could leave you with a big brown mess.

WHO IS A SOUND BATH GOOD FOR?

A sound bath is good for the time-pressed and super-stressed. Sound healing can help you if you suffer from the white-knuckle grip of anxiety or if you are so stressed and overwhelmed that you do not have time to learn a new skill. A sound bath is helpful for those who want to calm their inner chatter and let their mind switch off or for those who want to feel rested but have found that traditional meditation has not worked for them.

The type of people who come through my doors for a sound bath vary greatly. For that, I am grateful. For many years, sound healing was only accessible at the back of a dusty crystal shop and up some rickety steps. Now you can walk into your local gym, community centre or even national monument or gallery to experience the power of sound.

With the increased visibility, I have been able to see a wide variety of people immerse themselves in the benefits of therapeutic sound. These include new mothers who are sleep-deprived and need space held for them, CEOs making tough calls, entrepreneurs needing a boost of creativity, groups of friends who want to add some zen to their week, and couples who want to spend quality time together in a different way. There is no demographic of people who could not benefit from sound.

Sound healing can also help you when you are well and when things are going well. So often we neglect our deep inner needs when things are on the up, but some of the most meaningful work, especially in one-to-one sessions, is best done on these occasions. When you are feeling good and not actively trying to avert a crisis is an ideal time to check in with yourself. In fact, it will help to make sure you keep it that way.

This is how I see my osteopath. Carrying all my kit and crouching over playing bowls means I have to take care of my posture. Waiting until I am injured to get help will of course benefit me, but regular maintenance of my body (in this case) ensures that I can hopefully avoid a big issue. Sound healing can be used in the same way.

When things are good, sound healing can make them better. In the Western world, sometimes the way we look at health and wellbeing is from a position of the minimum that can be done, rather than the optimum we could achieve. I think that is in our collective psyche, too. The opposite of having chronic anxiety and feeling dangerously stressed is not feeling "less anxious than normal" and "just a bit" stressed, and things do not have to be "okay" when "great" is an option.

You could compare this to the recommended daily allowance (RDA) for vitamins. RDA is not designed for optimal health but is intended to give you the absolute minimum to survive. Mental health is often viewed with the same lens. You do not have to be desperately unhappy or harming yourself until you are allowed to want things to improve. You can choose a life where you are thriving even when things are going well. You can let go of any baggage that no longer warrants being in your life even if it does not require your urgent attention at that moment.

When you are feeling well and your nervous system is calm and well regulated, this is an excellent place to do deeper work to release old traumas and address issues.

It is important to remember that your energy is finite. You only have a certain amount you can expend in any given time. One of the reasons I tell clients to be mindful about worrying too much is that if you spend your energy worrying before something happens, you will have less energy to deal with it if it does happen. If you are in a constant state of anxiety and fear, you will not be equipped to handle it if something goes wrong. But if you work to build up your resilience, not only are you *not* worrying, but you are raising your baseline to cope in case something happens that warrants a stress response.

Working with your subconscious mind from a place of rest and happiness can also amplify other objectives. Want to manifest something? Chase your goals more efficiently? Perhaps you wish to reframe any negative thought patterns that have held you back in the past? The best time to do this is when you are not presently dealing with the consequences of it. Truly, there is no better time to do self-development work than when you are not in crisis mode.

YOUR BRAIN AS A BUSY OFFICE

This is my favourite analogy when it comes to describing how sound works and why it is good for you. Imagine your brain is like a busy office space with a medley of different filing cabinets: some are as tall as mountains, some are as wide as oceans and others as deep as a regular filing cabinet. There are also large empty desks in this office.

Each day, new files of different sizes and degrees of importance arrive and, due to the scale of this office, putting away these files takes time. Soon more and more files come in and begin to accumulate and there is no time to put them all away. So you are now putting some files on the previously empty desks. As even more files come in, they start arriving in a messy state, too. They are out of order, some even have pages missing and others have pages spilling out of them.

The desk surfaces fill up quickly with folders so some files now need to be stored on the floor. At first the office is semi-tidy, but soon you are tripping over those files and there is paper falling out of them. The files stacked high on the tables start to topple, so some of those are on the floor now as well. The folders you do manage to put away are filed so haphazardly that some end up in the wrong cabinets. You do not have time to shut the cabinet doors as more files keep coming in, so many of them are now left flung open. The stacks of papers and files on the floor mean the cabinet doors on the lower levels can't be opened either.

Your office is a mess. You are overwhelmed as the disorder seems unfixable. You start spending more hours at the office trying to crawl back a semblance of order until eventually it feels like you never go home. You are in the office all the time and yet it is still a mess.

This is your brain when you are stressed; the folders are your memories, experiences and feelings. When you are well rested and your brain spends time in a dream-like state, it is able to put these "files" away, categorizing them into the correct filing cabinet. If you ever need to retrieve one of these files – also known as experiences or emotions – you know where they are. Your brain puts everything away each day so that the desk space and the floors are clear – and this system works.

When a stressful event happens, there is an influx of files. When you are anxious, the files are in another language and need to be translated first before you can put them away. When a traumatic event happens, imagine a dump truck has unloaded a whole bunch of files into your office.

When you are chronically sleep-deprived and rest-deficient, the files come in at a rate whereby you can't put them away – resulting in the aforementioned mess. In real life, this translates to us becoming increasing short-fused and ill-tempered. Confusion or anger happens because the files are in the wrong place – this is why when you are stressed you may find yourself triggered by something meaningless.

During a sound bath, your brain is able to access that dream state, giving your brain time to put everything away and sort through the metaphorical files and folders. This is why, when some clients come to see me, they are shocked that a seemingly random, old memory comes up for them.

How stressed you are to start with will dictate how much you will process in a session. While sound healing is a shortcut to accessing a deeply restorative space, it is not a shortcut to everything. One sound bath can leave you feeling calmer, more rested, recharged and even reenergised, but it will not heal you.

If your brain has been a messy office for months or even years, you may find it difficult to switch off in your first sound bath. That does not mean it is not working or helping you. I am not here to mislead you, but to reassure you that if you go back and try again your experience will be worth it.

THE SUITABILITY OF SOUND HEALING

One of the best things about sound healing is that it is suitable for almost everyone. It does not matter which language you speak, which life experiences you have had, which tax bracket you are in. Sound is universal. Another huge plus is that you do not have to do anything either. When you attend a sound bath, you just bring yourself; you do not need to learn a skill and no active participation is required.

I am a huge fan of meditation, but learning to meditate can take years of dedication. With sound healing, the hard work is done for you: the skill set is with your practitioner, and your job is simply to listen.

How to Listen During a Sound Bath

While I like to conduct my sessions with as little outside noise around me, sometimes this is just not possible. An ambulance may go past, a car alarm may set off somewhere, and these things can't be predicted or helped. Some attendees have become agitated and annoyed at this. Afterwards, some may even complain about the general street noise they have heard. But environmental sounds are a part of life, and unless you are having a sound bath in a fully sound-proofed room, it is inevitable. Is this important? Does it impact the quality of your session? Does it "ruin" it? No, it does not. In fact, there is a lesson to be learnt within it.

As we will explore more deeply in Chapter 3, one of the biggest benefits of sound therapy and sound healing is learning to embrace more resonance in your life. Resistance will always be there. Toxic positivity teaches us that if you work hard enough you will be able to live a life free of resistance, that life can be just sunshine and daisies. Well, weeds are part of the package, too.

When an ambulance screeches past, think of it symbolically – that shrill sound from the ambulance siren represents all the things in life that happen outside of your say-so. You can't control it, and if you are in my session I can't control it either. You can't pre-empt it or prepare for it; it just happens. The magic lies in guiding yourself back to a state of resonance. The quicker you are able to do that, the more your mental wellbeing increases.

As I recommend to all my sound bath attendees, if you hear an unexpected sound and get distracted by it, first remove the guilt or annoyance that it distracted you, then simply guide yourself back to the sounds of the instruments. Refocus on what is being played. Getting frustrated and annoyed at the ambulance/loud noises/traffic is wasted energy and does not serve you. You do not benefit from keeping your focus on it. If anything, it will take you further away from a state of relaxation and back into the conscious mind, which is the opposite of what we are trying to achieve.

This is an important metaphor for life, too. If you spend all your time focusing on what you did not want to hear, about what has gone wrong or not worked out in the way you wanted it to, you will miss the good you *can* hear and see.

So, just like your thoughts, allow any additional environmental sounds to come, acknowledge them and let them go. Let them become part of the practice and then permit your mind to go back to the instrumental sounds you can hear. By returning your attention, you will sink back into the meditation, and at the end of the session you may not even remember the interruption. The more you can let go like this, the more you may find that your brain uses any sonic interruptions as part of the mind's eye journey.

As I've mentioned before, I end each of my sessions with a therapeutic percussion sequence. This involves using different chimes and shakers to gently bring clients back into the space and out of an ASC. I mention this here because when you let go and truly start to listen, you have the ability to sink very deep into a meditative state.

Some of my clients have not heard the transition from bowls to chimes and some have even missed the initial rousing shakes and trickle of a rainstick. They return back into the space only when the silence comes at the very end or they hear my voice. Only a handful of people have ever walked out from one of my group sound baths, and it has always been because someone very close to them was snoring. This usually indicates that they would have benefited the most from staying and trying to conquer their reaction. Not being able to tune out annoying background noises – or finding these sounds extremely infuriating – is a sign of nervous system dysregulation. Instead, try to sit with the discomfort and embrace the fact that it is trying to teach you something.

I always say that you do not have to do anything in a sound bath, and this is true, but if you wish to, you can heighten your experience by starting to treat it like a practice. If environmental sounds distract you too much initially and you work at it, you can get to a stage where a fire engine will go past and you just will not hear it. No, let me correct that – you will hear it, but not in a conscious way, and certainly not in a way that will pull you out of an ASC state. It will become part of your soundscape, and chances are you either will not remember it when the session ends, or it will be of such little consequence that you will not mention it or consciously register it.

Of course, the one exception is your phone. Everyone is so conditioned by all their modern tech that even in deep sleep they recognise the familiar buzz of their phone. This will whip you out of your meditation, so turn it off before every session.

Listening With Intention

Intention is crucial when listening in a sound bath. Sound healing pioneer and musician Jonathan Goldman first devised the formula: Frequency + Intent = Healing. And he was right. The intention you bring into a sound bath will also determine what you receive from it.

Intention is the secret ingredient, the glue that binds everything together. So be open on your journey to sound healing and acknowledge that any possibility is available to you when you lie down to be bathed in sound. Keep your heart open, as well as your ears, during a session. Do not go to a sound bath with expectations but do leave with gratitude for what happens.

You can have some incredible experiences with sound healing, but remember that some of these come with time and building up a practice. You may have one, two or even three sessions before experiencing certain benefits mentioned in this book. Just be open to what you are there to experience and give yourself time for it to work. You do not get abs from eating one salad so you can't expect one sound healing session to eradicate the consequences of not resting for years in one go.

ADDRESSING COMMON QUESTIONS ABOUT SOUND HEALING

Whenever you try something for the first time, it is natural to have some questions. I am asked a handful of the same things over and over again, so I am going to delve into them here. If you are in a session and something happens that you have not experienced before, do not be afraid to ask your practitioner what is happening or why it happened.

How Often Can I Have Sound Healing?

You would be surprised at how many people ask this, and truth be told there is no correct answer. Come as often as you need. There is no downside. This is not like exercising where you are at risk of wearing yourself out, and it is not like talking therapy where you need time to process what you have discussed. Spending time feeling rested is never a bad thing. After all, we are supposed to be deeply resting every night for a minimum of eight hours.

It is also worth remembering that while you are enjoying the sound, so am I. I teach back-to-back classes and have back-to-back private clients – so while you may not need two sessions a day, there would be nothing wrong with you having them. I usually recommend once a week to help you build up a good practice and twice a week for whenever you need a bit of extra support. But if you wanted to listen to a sound bath each night before bed, that is completely fine, too.

What If I Fall Asleep?

If I had a pound for every time someone apologized or pre-warned me that they might fall asleep in a sound bath, I would have the purchasing power to buy all the instruments. Firstly, rest needs a rebrand. It is okay to rest for the sake of rest; this is not being lazy, nor is it "pointless" or a "waste of time". Culturally, our collective sense of productivity is attached to everything, even a sound bath, where the goal is simply to be present. Everything else that happens is a bonus. You do not have to turbo-boost your rest to make it "worthwhile".

If you fall asleep in a sound bath, that is okay. Prefacing a session with a disclaimer that you may end up in the land of Zzzz's feels like you are trying to absolve yourself of any guilt if you do so. But guilt and shame do not have a place in sound meditation. If sleep is what you need the most in that moment, fall asleep. We will explore the different types of rest in Chapter 3, but I really do believe that if your physical need demands priority, then give it priority.

Do not spend the session fighting against it. Sleep can also signify safety. If you struggle to fall or stay asleep at home but quickly nod off in class, then think about why this could be. Perhaps your bedroom needs some adjustments to make it more sleep-friendly?

When you attend a sound bath, you are not just hearing the sounds, you are also feeling their vibrations and therefore receiving the benefits that come with this. The sound bath will still be working if you are asleep. Sure, you may skip the healing phase of theta brainwave activity, but delta dominance (which happens in deep sleep) also serves a purpose and enables your physical body to regenerate.

Finally, you could be in the deepest sleep of your life but your practitioner is going to keep playing – we do not sneak off to have a cup of chai.

Is It Normal If I . . .

Cry, twitch, shudder, shake, sigh heavily . . . yes, it is! While so much can happen to us mentally or emotionally during a sound bath, we can also have a variety of physical reactions. You may feel so emotionally moved that you want to cry. If this is the case – go with it. Let your body release in the ways it wants to. Holding in your tears or being worried that you may cry puts you back into your thinking brain and your physical body and therefore stops you from being able to reach a higher level of meditation. Remember that there is no right or wrong way to surrender; the key is simply to give yourself permission to do this.

There is no shame or judgement if your body chooses its release through tears. Shedding tears is cathartic, a self-soothing act and a form of self-regulation. Tears serve a multitude of purposes. Crying flushes the body of stress hormones[15] while releasing oxytocin and endorphins to ease pain – be it emotional or physical.

Emotional tears (as opposed to lubricating or irritant tears) also contain more protein, so they fall down your cheeks more slowly. Scientists believe this is so the tears can be seen by your community, which in turn gives them the opportunity to then care for you. If I or your practitioner see your tears in a sound bath it is because we are meant to. In that moment, we are your community and are holding space for you to release in a caring environment.

Just like crying, sighing loudly, muscle spasms and jolts are also examples of your body experiencing release. This time the difference is that you are letting go of the stored-up stress and tension you have been holding on to. I see clients do full-body shudders, flick their feet or twitch their arms. This is all perfectly normal.

Your central nervous system is constantly sending signals to your muscles telling them to expand or contract. When you are lying down and resting, your muscle fibres may have extra energy they need to get rid of so they can relax with your brain – cue the twitching. If you have been in fight-or-flight mode – i.e. highly stressed and anxious – for a while or before you engaged in your sound bath, this reaction could be more intense.

This is also the reason that, if someone comes into a sound bath visibly stressed or the collective energy feels extremely high in a group session, I have been known to make everyone channel their inner Taylor Swift and "shake it off". Shaking gets rid of some of that excess energy and helps to recalibrate the nervous system, in turn calming the body. This follows the same principle as TRE (Trauma Release Exercise). In nature, we see animals behaving like this all the time – whether it is the gazelle or a stripy zebra; if they have escaped an attack by a predator, they will shake to release tension.

Dancing is also a great way to calm yourself. If you have had a particularly intense few days or even weeks and feel like you need to let off some steam before attending a sound bath, or before listening to the audio I have recorded for you (see page 239), put on an upbeat song and shake out your body first.

Rest and . . . Digest

Another way your body indicates that it is in a parasympathetic state is that your tummy starts to rumble. People get so self-conscious about this, but, again – please do not worry about it. It sounds louder to you than to me or anyone who may be in the room. Also, sonically speaking, it is a low rumbling sound and therefore low in pitch, so it is unlikely to wake anyone from a state of rest even if they did hear it.

Your tummy rumbling when you are relaxing is perfectly normal. After all, it is called rest and digest for a reason. When you are stressed and anxious and in a fight-or-flight state, blood is diverted from non-essential systems such as digestion to your heart, lungs and limbs. Your breathing gets faster and your heart beats more quickly so you can react to the impending danger.

If you are chronically stressed, you are likely to have gut issues and experience bloating. Your body thinks you are in danger and can't prioritize healthy digestion. When you are relaxing, your blood can be redirected to the digestive system, so your stomach starts to rumble. Blood circulation also increases in the brain and other organs, moving away from your limbs – which is why you tend to feel cold in a sound bath or any type of deep meditative work.

Can I Come to a Sound Bath if I'm Hard of Hearing or Deaf?

Yes, you can. Because you also feel the vibrations in an in-person sound bath, many clients who are hard of hearing or wear hearing aids actually find attending a sound bath helps them to relax the strained muscles around their ears and necks.

If you have tinnitus, speak to your practitioner before booking a session. There are essentially two types of tinnitus; depending on which type you have and how it occurred, sound could potentially help to relax you and give you a break from the ringing that happens in your ear. If your tinnitus is the result of an ear injury, or hearing loss from exposure to continuous loud sounds, unfortunately a sound bath is likely to irritate your ears further.

The inner ear is lined with tiny little hairs called cilia. Different sections of cilia pick up different frequency bands. At the base of each of these hairs is a receptor. The sound that passes over the hair sends a message to this receptor, which then delivers a signal to the brain to interpret what you are hearing. Unfortunately, if these little hairs get damaged, there is not much you can do. But if your tinnitus is stress-related and comes on when you are highly anxious, sound can help to destress you and therefore gently ease the symptoms you are suffering from. Even if this is the case, please tell your practitioner about your tinnitus.

The same applies if you are neurodiverse or suffer from any sound sensitivities. Inform your practitioner so that they can decide on the correct course to follow. The solution could be as easy as deciding where you sit in a group session or giving you extra advice on what to do if you become uncomfortable at any time. Equally, this reassures the practitioner. In a sound bath, a good practitioner will be scanning the room the entire time to keep an eye on everyone. If you are harmlessly stimming, that is absolutely fine, but if we do not know this is what you are doing, we may be concerned that the sounds are not encouraging you to rest.

I have seen potential clients who only mention in passing that they have a pacemaker and do not include it on their treatment consent forms. It is very important to tell the practitioner in advance because while sound healing is suitable for nearly everyone, in-person sessions are not suitable for people who wear pacemakers, because the vibrations from the instruments can interfere with the way the device operates.

What Should I Wear?

We have already discussed that you will not be getting wet during a sound bath, so swimsuits can be left at home! Comfort is key, so wear whatever you feel comfortable in. If you are having a sound meditation in the middle of the work day and are in an office-friendly suit or dress, that is fine. You do not need to put yoga or workout clothes on. If you are winding down and will be heading to bed straight after your session, wear your cosiest tracksuit. If you are having a private session at home, wear your pyjamas if you wish.

The only time you need to be mindful of your clothing is if you are having a vibroacoustic massage. If that is the case, avoid bulky waistbands, belts and hard fabrics such as denim and anything with zips and metal fastenings that could vibrate against a bowl.

SOUND: HEALING, THERAPY AND MASSAGE

Sound can be used in many ways. As I have mentioned, a sound bath is simply the name we give to a session that involves therapeutic sound. While this book focuses on sound healing, it is important to know that other options are also available.

Sound Healing Versus Sound Therapy

While sound healing and sound therapy are used interchangeably (hands up, I am guilty of this, too – especially in the beginning when I was trying to get the message out there about sound meditation), I am eager for everyone to understand that there is a difference between them.

Sound healing can happen in both group and individual settings and any therapeutic-grade instruments can be used (we will explore this in more detail later). The main aim is to get you into a place of rest, slow down your heart and breath rate and calm your central nervous system. Over time, we also want to increase your vagal tone (see Chapter 3). Using sound in this way opens up a whole host of other tools to enrich your life – all of which we will be exploring in this book.

Sound therapy is exactly that: sound + therapy. Practised in a one-to-one setting, it combines the best of sound healing with a self-reflective practice to help you dive deeper. Sessions begin and end with talking. You get to explore and choose what you want to address; the sound becomes a container within which you can work through and conquer the issue at hand. This is not psychotherapy, so the talking is limited and it is done in a more abstract way than just talking normally.

Sonically speaking, sound therapy also uses a combination of pure sounds, harmonics, different frequencies and specialist instruments played in a therapeutic way, similar to sound healing. But because the sessions are one-to-one and bespoke to your particular concerns, the sound is tailored exactly to what the practitioner thinks you need. This means that at times the sound can be confronting, stimulating or even challenging, and dissonant harmonies will appear. Sound does not always have to be harmonious to be healing. This is especially true in sound therapy. The discomfort is deliberately placed there to get you an answer, to shift your inner dialogue and to dislodge your self-limiting thought processes.

While you will still get to a place of rest, with a sound therapy session you may not always get to a place of relaxation. This does not mean it is not a successful or worthwhile sound experience. On the contrary, it may be exactly what you needed for long-term growth and healing. I liken it to when you have a traditional, physical massage. A therapist may find a knot in your shoulder and apply some gentle pressure there. It may not feel the best in the exact moment they do that, but the release that follows always hits the spot. Sound can also be used in the same way. Practitioners can find the energetic knots and emotional blockages and sonically apply some gentle pressure.

Vibroacoustic Massage

Designed 40 years ago by Estonian physician Professor Saima Tamm, vibroacoustic massage is another fantastic way to use sound to deeply relax you. Using either Himalayan bowls or frame drums, the low-frequency sound and physical vibrations from these instruments help to ease any tension you are holding in your muscles.

You remain fully clothed while the Himalayan bowls are placed on your body. The frame drum does not touch you but is played close enough for you to feel the vibrations coming from it.

Drum Circles

Also called drum journeys, drum circles are sessions where the main instrument used in a sound healing session is a drum. Some drum circles require active participation where you will also be drumming while in others you will lie down to just receive the sound instead.

Soundscapes

A soundscape is a broad term used to describe the environment you are in. Like a landscape but in audio form, it is everything you can hear. Your natural soundscape may include the hum of traffic, planes flying over, the playing of children at a nearby school, the gentle ebb and flow of a river or even swaying trees.

In sound therapy, the soundscape we talk about is manmade and therapeutic. A practitioner will ask you to choose an instrument (such as a shaker) and instruct you to play it in a way that expresses a particular emotion you are going through. This makes soundscapes great tools of transformation and enables a client to express how they are feeling when they do not have the words for it.

CASE STUDY: A WHOLE NEW WOMAN

Whenever I am asked about the difference sound can make, my mind always returns to one particular client. I will never forget the first time she came to see me. Winter was fast approaching, the nights were cold and stormy and her friend (another client) had mentioned to her that this type of work, sound, could be the one that finally works.

As a practitioner you should always keep some boundaries with your time but I let our first session significantly overrun. I had to help her – she was frazzled and very much in a freeze response. The overwhelm was consuming her and I could see where it could quickly fall into hopelessness. She was not sleeping well either and that was inevitably making everything 100 times worse.

I decided to get her to draw before the sound bath. (I do this with certain clients and we will explore how you can do this in Chapter 7). As she handed me the image back, I knew she needed to be here.

I chose the warmth of the Himalayan singing bowls for her and insisted she come back each week on the same day, which she did. Each week, she drew and I played the bowls and I even gave her some homework by way of the exercises I cover in Chapter 6.

The woman who came in on her fifth session was not the same person at the start of the month. Her hair, skin and make-up were all different. I could not be sure if her clothes were different or if they just appeared that way because of how she was standing; she was holding herself differently and seemed taller. She was smiling and there was a general ease in her demeanour. Work had improved and things just seemed easier than before, she enthusiastically told me as she sat down.

I handed her a piece of paper and my pot of colouring pencils and asked her to draw, just as I had in our four previous sessions. She picked different colours and her pencil strokes were softer, kinder and gentler on the paper. The colourful waxy pigments glided across the page; unlike in week one, where they left grooves on the page below.

With the work we had done over the weeks, her mind had become clearer and her body had become calmer. Sound had allowed her to decompress, to shift her perspective and shed the weight of anxiety that had been suffocating her so fiercely. She could not believe how much she resonated with the sounds of the singing bowls and how, as the weeks went on, she relaxed more quickly. She explained that it was as though the warmth of each strike of the bowl had somehow glued her back together.

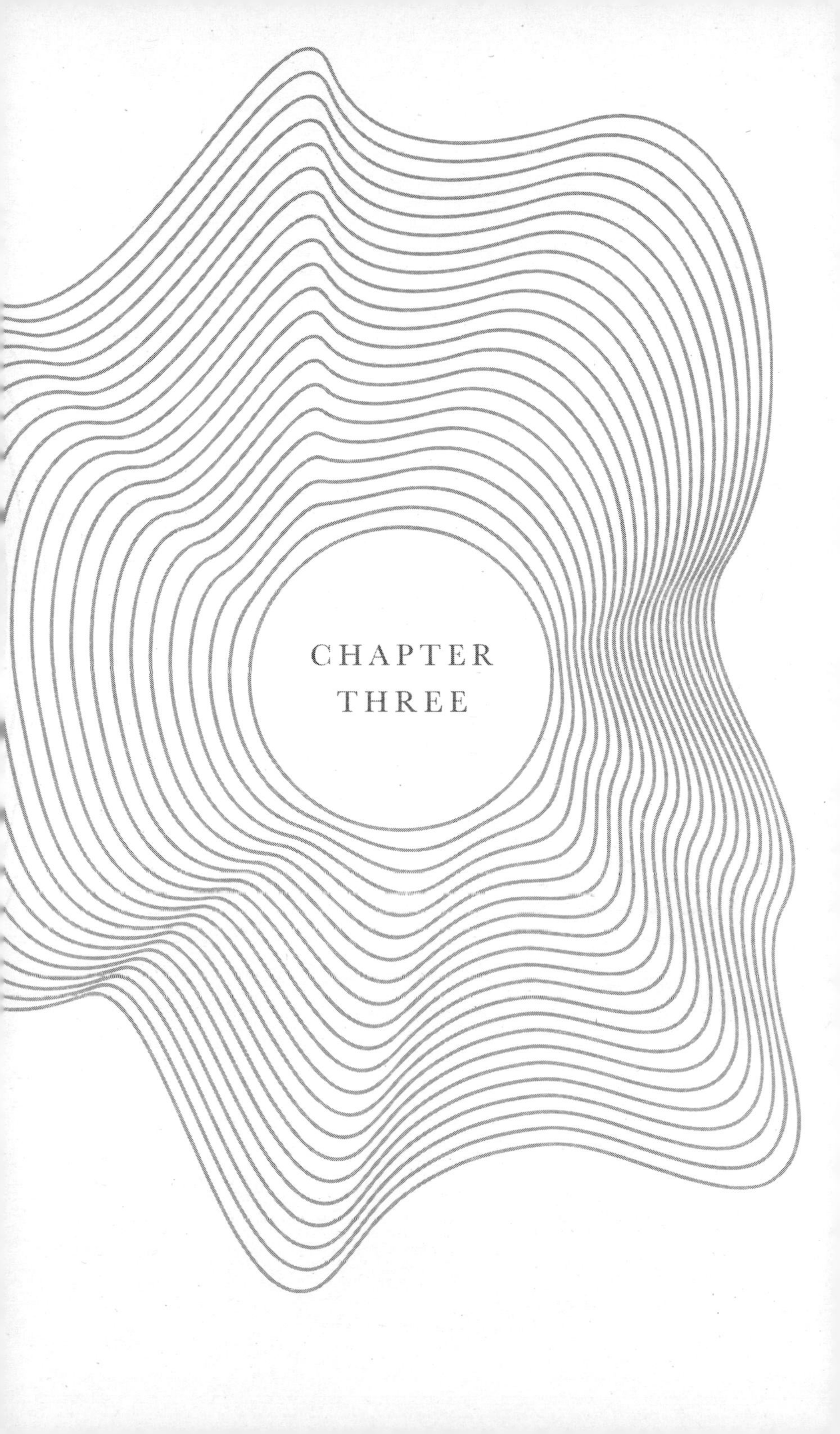

CHAPTER THREE

"Sound will be the medicine of the future."

– Edgar Cayce

How We Heal Through Sound

Hustle culture is the concept that work is everything, burnout is a badge of honour and rest does not matter. It has gripped most of us for many years now, but the tide finally seems to be turning as more people realise that the "5am Club" is not only unsustainable, but can be harmful to our health.

According to a study by global consulting firm Deloitte, 77 per cent of people have experienced burnout at their job.[16] Most people in this study admitted to dealing with feelings of disengagement, lacking both energy and motivation from having experienced high levels of work-related stress.[17]

Also known as grind culture, hustling is when you focus on working long hours and never rest. The idea behind this is that working hard trumps everything. However, the research proves otherwise. Being a work martyr is not only bad for your brain and body but it actually does not achieve anything. One study from Stanford University found that not only does productivity decline sharply after 50 hours of work, but it completely drops off after 55 hours.[18]

Sound healing is your antidote to hustle culture. It provides the mental clarity to work smarter rather than harder. Sound boosts your wellbeing and helps restart your focus – it clears the noise. Therapeutic sound allows you to heal by letting you rest deeply and recharge effortlessly. And it does this in a variety of ways.

THE CONCEPT OF REST

Before exploring all the ways that sound healing helps you rest, ask yourself: "What does rest mean to me?" By this, I mean what does rest mean to *you* personally and what does it look like? Have a think, and maybe even grab a piece of paper and jot down your thoughts. When you have finished, take a close look at your answers. Is rest simply the absence of activity? Is it the opposite to being busy? Perhaps it is not moving or not working? Is rest the absence of feeling stressed? Is it being on holiday or the time you give yourself after having done all your chores and ticked everything off your to-do lists? Is rest lying on the sofa and watching Netflix, or sitting up in bed scrolling through Instagram or TikTok?

According to the dictionary, rest seems to be something you earn: "A period of relaxing, sleeping, or doing nothing after a period of activity." Or rest is an absence of doing: "Freedom from activity or labour: a state of motionlessness or inactivity."

Therapeutically speaking, neither of these definitions is correct. Rest is about recovery and about replenishing and recharging your batteries, but it is not a treat or a luxury. You do not, as the dictionary suggests, have to *earn* it by working extremely hard first. Rest can be anything that increases your mental and physical wellbeing. It involves a restorative practice to bring you back from a place of overwhelm and exhaustion. Rest is a fundamental necessity, required and non-negotiable for health, wealth and happiness.

Most importantly, rest is needed *before* it becomes necessary. Yet most of us are not getting enough of it, having been almost brainwashed to keep going. We embraced the narrative that life is all about hustle and grind culture, the #girlboss aesthetic, and "sleep when you are dead" sayings. This lifestyle is unsustainable – and I should know, as I was an active participant in it.

As both the Stanford and Deloitte studies show, when you do not give yourself access to adequate rest you leave yourself vulnerable to chronic stress, overwhelm, anxiety and burnout, all of which can seep their toxicity into every aspect of your life. The absence of rest impacts every part of you. Your emotional intelligence is likely to decrease – leaving you more prone to being irritable with a short fuse, fracturing personal and professional relationships as a result. You shift into a resistance-heavy outlook, expecting the worst to happen, looking for things to go wrong and for disappointment. You may withdraw from family or friends – simply not having the bandwidth to either ask for support or to maintain those ties.

Physically, you may be exhausted and running on empty, unable to fully switch off but in desperate need of rest. Your sleep may become interrupted and in short supply. Without adequate rest, you are at risk of a plethora of diseases, too. All of this compounds into your mental health – feeling blue and unhappy, incredibly unsatisfied with life, lacking focus and losing your creativity and problem-solving abilities.

Sound healing has the ability to meet you at your most exhausted, most dissatisfied and most burnt out and wrap you in its cocoon of soothing, restful waves. This is so important because being fully, adequately and nourishingly rested helps you *feel* better mentally, improving your mood and therefore your outlook. Your body has time to repair, your energy is replenished, and you can return to a state of homoeostasis. Rest lets you reduce your stress levels, eases your anxiety and helps you build better connections.

The Seven Types of Rest

Gone are the days when lying on your sofa doomscrolling through Netflix was classed as rest. These days we know that rest is much more complex and multi-faceted.

In her book, *Scared Rest*, Dr Saundra Dalton-Smith identified seven types of rest: physical, mental, emotional, social, sensory, spiritual and creative.[19] Sound healing manages, in my opinion, to achieve each of these types in different ways.

- **Physical** – a sound bath provides time to "just be" without any physical exertion while relaxing your muscles.
- **Mental** – sound healing provides an opportunity to quieten your mind and not engage in problem-solving and concentration.
- **Emotional** – sound healing regulates your nervous system; a happy, well-functioning nervous system equals better emotional intelligence.
- **Social** – attending a sound bath puts you into an environment where you are likely to make supportive social connections.
- **Sensory** – with your eyes closed, you can have valuable time away from screens, lights and noises and envelop yourself in deeply harmonious sounds instead.

- **Spiritual** – having an Altered State of Consciousness (ASC) experience can connect you to your inner wisdom and knowledge.
- **Creative** – being in a deep meditative state as experienced within a sound bath allows you to tap into your creativity and purpose.

YOUR BRAIN'S NATURAL RHYTHM

To understand how sound healing engages you in deep rest and allows for so many forms of healing to take place, let me explain a bit more about how your brain works. Your brain is made up of a huge number of different cells, including a few billion neurons. Neurons communicate information throughout your grey matter and, in doing so, create an electrical signature.

Depending on which activity you are doing and how your brain is responding to that task, these signatures can be grouped together by rhythms. These rhythms have different speeds that make up your brainwaves. Contrary to popular belief, your brain does not just have one brainwave type at a time. Different parts of the brain have been shown to produce different brainwaves simultaneously. However, there is always a dominant brainwave and this is what we will be focusing on.

The five main types of brainwaves – delta, theta, alpha, beta and gamma – all serve different purposes.

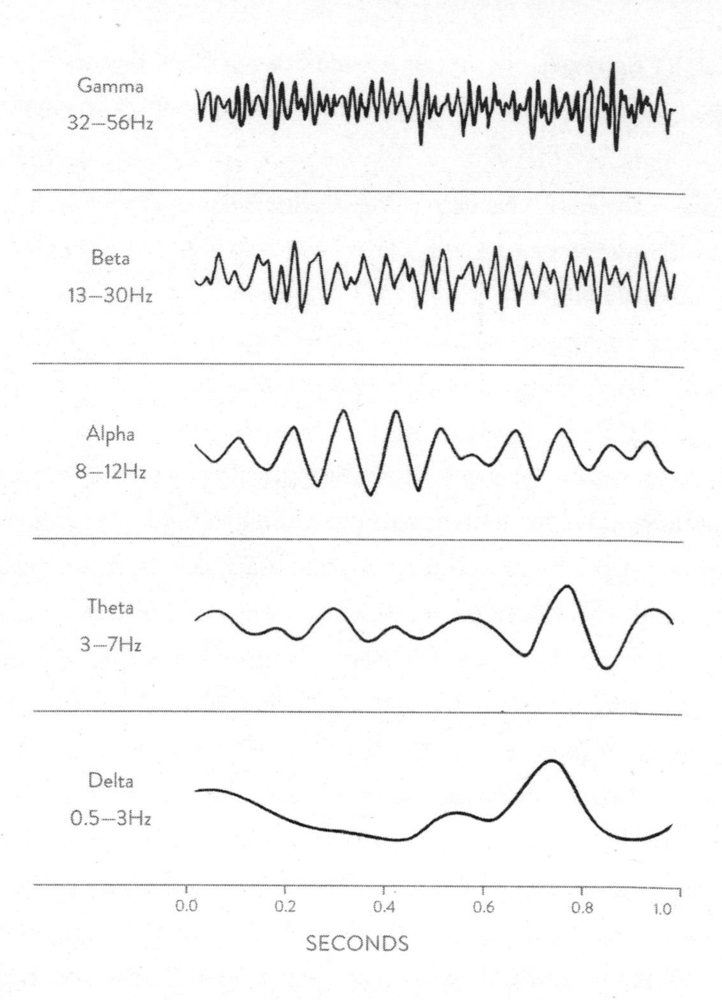

Gamma (30–56Hz)

Initially dismissed as just a background brainwave, gamma is the fastest brainwave state and is usually seen in hyperawareness. This brainwave activity is associated with higher levels of insight and information, expanded consciousness and self-awareness and even out-of-body experiences. It occurs when your brain is simultaneously processing information from different brain areas – unifying all the information it is receiving.

Beta (13–30Hz)

Beta waves have a faster and regular frequency that uses a lot of energy. These are dominant when you are awake, alert and concentrating – your day-to-day brainwave state. You are in beta dominance when processing new information, solving problems, making decisions and generally having complex thoughts or thought processes. Stimulants such as coffee or nicotine can move your brain into a beta state, similar to how meditation can bring on an alpha-dominant brainwave state.

Too much time spent in a beta brainwave state is detrimental because this is when you become stressed and anxious. It throws off the optimal brain chemical levels needed for proper brain function, leaving you mentally and physically fatigued.

Alpha (8–12Hz)

Alpha is a more uniform, synchronized brainwave that is slightly faster than theta. This is the brainwave of relaxation and the one we speak about most in meditation, yoga nidra or anything that is designed to bring you back to a place of calm and peace. Alpha is the brainwave of daydreaming or taking a relaxing soak. Your mind is conscious and awake, but your body is in deep relaxation. You feel good, you are at peace and things are well.

Theta (3–7Hz)

Theta brainwaves are also slow and irregular, but these large waves have a slightly higher frequency than delta waves. Theta-dominant brainwave activity happens in near-sleep. This is when you are dreaming or in the REM (rapid eye movement) part of your sleep cycle.

Deep meditation also induces this state. When your brain is in theta dominance, the mind processes your thoughts and emotions. This is the gateway to your subconscious, too – where your true hopes and fears live. Accessing this space also allows you to tap into your intuition and creativity.

Because your brain produces complex imagery in this state, not only do dreams occur here, but mystical or spiritual experiences are associated with this state as well. Essential brain chemicals are balanced here, too, which is why after being in a theta state you feel mentally refreshed.

Delta (0.5–3Hz)

Delta brainwaves are slow, large waves with a low frequency. You experience these in deep, dreamless restorative sleep. When you are in this state, your body heals and repairs itself and your brain has a rest. If you do not get enough deep delta sleep, you are more prone to getting ill.

SOUND HEALING AND BRAIN-WAVE ACTIVITY

Meditation takes you from a stressed and anxious beta-dominant brainwave state and puts you into a relaxed alpha-dominant brainwave state. While extremely worthwhile, this can take many years to master. Sound healing works by providing a shortcut to this process.

The instruments used in sound healing recreate these slower frequencies. Then through a process called auditory driving and sympathetic resonance – which basically means that your brainwaves change to match the frequency of the sounds you are hearing – your brainwaves slow down and sync into that restorative and dreamy alpha-dominant brainwave state. Not only does this require little effort from you, but because it is an automatic response by your brain to whatever you are listening to, it just happens. This is why sound healing is an attractive option for someone who has always struggled with regular forms of meditation and finds it hard to switch off that "monkey brain" chatter.

When your brain syncs to a slower wave pattern, this tells your body that you are safe, you can relax and you do not need to be concentrating or on high alert. As a result, your heart rate starts to slow down, your breathing gets slower and softer, and your blood pressure drops. Your parasympathetic system is activated. You are in "rest and digest", which is the opposite to the fight-or-flight response. You are calm and become deeply relaxed, your overall sense of wellbeing improves and your mood is lifted – you do not feel as stressed as you were.

Sleep experts I have interviewed over the years have often stated that problems relating to sleep and insomnia are linked to an inability to relax rather than an inability to sleep. The longer you spend in the slower brainwave states of relaxation, the easier it will become to switch off before bedtime, especially if you are listening to a sound healing track beforehand.

Similar to climbing a series of steps, your brain needs to work through the different brainwaves states to get to the prize of deep sleep delta waves. If you are helping your brain navigate itself from beta waves to alpha waves and then to theta waves, delta is only one step between you and the land of

Zzzz's. It will be much easier for your body to accomplish this progression when you are relaxed before bedtime than when you are stressed and trying to get to sleep. Slowly you will find that both your sleep quality and sleep quantity improve.

Even as a sound therapist, I can fully admit that sleep trumps everything. On occasion, clients who have been travelling and are jetlagged have fallen asleep at home and missed a class with me. My advice to them, and to you now, is what a personal trainer friend shared with me 20 years ago: when you have a choice between "X" and sleep, always choose sleep. That "X" could be sound healing, exercise, time in nature, socializing – all things that are good for you. But it is important to allow sleep to win. Sleep is needed for physical, mental and even emotional rest. Without good sleep and therefore physical rest, you can't access the other types of rest. Fortunately, sound healing happens to be an easy gateway into sleep.

EXPERIENCING AN ALTERED STATE OF CONSCIOUSNESS (ASC)

An Altered State of Consciousness (ASC) is the pinnacle of a sound bath – the juicy bit of the fruit. It is when clients describe the sound coming in as caramel and drenching them, or they suggest that I must have moved them in the room or turned on an array of different-hued lights. I know it has happened when a sound bath attendee warily reveals that they felt like they were levitating. ASC is the addictive part of the sound bath experience and is the part I love hearing about from those attending one of my sessions. ASC is also one of the ways in which sound healing provides mental rest.

Firstly, let me break down what an ASC is. It is a trance-like psychological state that is different to when you are awake, and usually happens when your brain is in either an alpha- or theta-dominant brainwave state. Having an ASC is good for both your brain health and overall health. When you are in an ASC, your brain has time to correct its internal mineral levels, repair cells and bring your nervous system back into balance. ASC is also when your brain processes imagery and stores your memories; while this may all sound complicated, your brain naturally does this all the time.

You do not need to be in deep meditation to experience an ASC. You may experience one when you are daydreaming, when you are looking out the window and you zone out, or are reading and drift off for a few minutes. The problem these days is that we are so connected all the time and our brains are constantly switched on. Think back to the last train journey or bus ride you took. Did you stare out the window and let your brain wander with its thoughts? Probably not. It is most likely that you were catching up on social media or checking emails or maybe even watching the latest episode of your favourite Netflix show.

We live in a world where people use caffeine and sugar to chemically refocus their concentration every time they feel themselves drifting off into a brain-friendly dip. In fact, I am doing this right now. My publishers may not let me keep this in here, but writing a book is one big Herculean effort in sustained concentration. Even in the moments when I would usually let my mind wander, I am having a sugary drink to snap me back into focus mode.

An ASC also happens when you are in REM sleep, so if you have sleep issues or insomnia you are at risk of missing this vital time. The trouble is that this downtime for your brain is not optional – it is essential.

Just as a wavelength has peaks and troughs, your concentration should do the same. The amount of time your brain is engaged in a beta-dominant brainwave state – where you are either concentrating, focused, engaged or thinking – should naturally be interjected with moments of daydreaming, of relaxation, where you slip into an alpha-dominant brainwave state. There should be an ebb and flow to your day, but modern lifestyles do not leave enough room for that natural rhythm.

Hands up if you reach for your smartphone the instant you wake up in the morning or if it is the last thing you look at before you go to sleep. There is no judgement on my part – I used to be guilty of this, too. So guilty that years before I became a sound therapist, there were countless times when I even fell asleep with my phone in my hand.

The phone is not the problem, as such, but it is our attachment to it. Email culture, 24-hour connectivity culture, hustle culture . . . Whatever you call it, it is a cultural problem, a societal issue that has made productivity the ultimate goal and rest the enemy.

A lack of downtime does not work for the human brain, which has evolved over millennia. Your brain needs time in an ASC state to function healthily. So, what happens when you're in this state? Your brain takes its own break. Have you ever driven home and can't remember turning into your road or even parking and locking the car? Maybe you have zoned out in the shower and can't remember whether or not you used conditioner or not. My personal favourite is leaving the house and then having to turn back and check if I have turned the hair straighteners off. These are all examples of your brain stealing some ASC time for itself.

When you meditate deeply or experience a sound bath, you allow your brain to stay in a lower brainwave frequency for a sustained period, providing ample time to tap into an ASC and benefit from it. The more time you spend letting your brain enjoy these lower frequencies, the stronger and longer your bouts of ASC can become. Or, at least, this is what seems to happen to clients with a regular sound healing practice. Within weeks of attending sessions consistently, clients progress from saying they could see some different-coloured lights to describing deeply complex imagery and rolling scenes of visuals. Unlike hallucinogenic drugs, which can also induce an ASC, there are no risks when you achieve it in a sound bath.

Your Unique Experience of ASC

Each sound bath, and each experience of that sound bath, is completely unique. Two people can be lying down in the same sound bath and have a vastly different experience, with certain sounds resonating more for one person than the other, and vice versa.

The ocean drum may remind one person of their favourite beachside holiday or of their home along the golden sandy shores of Australia, but it may be annoying for another person – signalling that the session is drawing to a close. I use this example because it has happened. The Australian client was so relaxed that when the oceanic rumble of the drum kicked in, she found herself transported back home. The sound lifted her up and enveloped her in a feeling of familiarity and comfort. The client next to her blinked her eyes open, eager to see what was going on and wondered what on earth I was doing to create the sounds of crashing waves in a central London location.

One client may not be able to get enough of the chime as it takes them back to a sacred memory of childhood where a ballerina danced within a jewellery box to a similar tune. For another client, that same chime wakes them up from their dreamy, sleep-like slumber.

Each sound bath experience is as individual as your DNA. Not only do your life experiences impact on how you receive the sounds within a sound bath, but your mood that day, what happens in the moments before you enter, your hearing, your stress levels and your brain will all play their part as well. How you enter an ASC state and which type of ASC experience you have are controlled by all those factors and more.

When you undergo an ASC, part of your brain becomes engaged. The ASC you experience will depend on how the different circuits in this part of your brain interact and are activated. This is why experiences differ so greatly in each session between clients, but also why you will not have the same experience from session to session.

POSITIVE SOUND BATH EXPERIENCES

On the whole, sound healing sessions are a positive experience. Why? Because the way a practitioner plays for a group is different to how they would play in a bespoke one-to-one therapeutic session.

During an individual treatment, the session is there to cater to your exact needs. This may sometimes mean playing dissonant chords and intervals, which can feel a touch uncomfortable – and it is meant to be. Those chords are supposed to challenge you, and your practitioner will be watching to see where they can take you sonically.

In a group session, rest is the primary focus, to soften the state you are in and to allow for gentle release. There may be some light resistance, but the main objective is to lull you into a state of ease. The session will be designed to let your brainwaves sync to the slower, harmonious sounds of the instruments, and to take you to a slower brainwave state where rest and creativity are unleashed. When you move from a thinking beta to a relaxed alpha to a slow, dreamy theta-dominant brainwave state, you may experience an ASC.

The ASC experience is as individual as your brain, but some feelings and sensations often come up for many people. I have listed some of the most common responses below.

- "I could see lots of different colours."
- "I couldn't fully feel my arms or legs."
- "I started to tingle."
- "I just felt complete joy, bliss."
- "No way was that an hour, I felt like I just put my head down for ten minutes."
- "Everything makes more sense now."

These experiences, while vastly different, can be loosely categorized under a few main branches. Here, I will explain the main positive experiences that crop up in sessions.

Positive Mood

Positive mood is pretty self-explanatory. You feel like a weight has been lifted off you, that life is good and things that annoyed you prior to the session just do not matter. You feel happy and you are at peace.

Complex Imagery

Complex imagery is when you see different colours and shapes and experience vivid images. You may feel like you are in a dream.

Insightfulness

As I have said previously, I am not a healer, and nor are most other practitioners like me. In sound work, we believe that *you* have the answers and the wisdom you need. Sound simply connects you to that wisdom. The insightfulness state of ASC is when you finish a session with a breakthrough, or a sense of deep clarity. Insightfulness helps you to see more than what appears to be there at first glance, enabling you to find a resolution, a solution and an answer – sometimes to a problem you did not even realize you had.

Oceanic Boundlessness

If you feel as though you are floating, tingling or not quite sure where your limbs are, you are in an oceanic boundless state. This is when it feels like your edges have disappeared and you are not sure where you end and your surroundings begin.

Transcendence of Time

I probably hear this one the most – when people can't believe that the session is over and that it is time to slowly awaken and stir from their place of rest. People often completely lose track of time and feel as if the session only lasted ten minutes or so. When you experience this form of ASC, it is as if time no longer runs the same way, that it has whizzed by. Some clients can also experience a transcendence of time in the opposite way, too, which means a session can feel longer, as if they have rested for several hours.

Spiritual Experience

Spiritualism can mean different things to different people. So, even if you do not consider yourself traditionally religious or spiritual, you can still experience this type of ASC state. The essence of this type of ASC is that you feel connected to a power higher and more profound than you.

Experience of Unity

This is when you experience bliss, when everything seems within your grasp and you feel connected to those around you. It is a flow state where every possibility could come to fruition. You may feel as though you are being wrapped up in a blanket of cosmic sound, a magical otherworldly enveloping of hope.

Because you access the deeper layers of your consciousness when you are in an ASC, experiencing it encourages a greater level of self-awareness and emotional intelligence no matter how it manifests. In fact, introspection and the ability to look inward and connect better to your inner feelings, noticing your bodily sensations and understanding them, helps with self-regulation. You can't control what you do not understand; to notice what you are feeling, you must recognize that you are feeling it.

VAGUS NERVE STIMULATION BENEFITS

If you are wondering whether you can benefit from sound healing without an ASC, the answer is yes. While an ASC state is a wonderful thing, you do not need to feel as though you are floating off into a different dimension on a rainbow of colours for sound healing to benefit you. When you are in a sound bath, you are not just listening to the sounds, you are feeling them, too. The vibrations created by the instruments are surrounding you and travelling through you. This makes sound healing an excellent tool for vagus nerve stimulation.

Your vagus nerve is the longest nerve in your body. It starts at the base of your brain, runs past the back of your ear (an important fact we will come back to) and travels down your entire body, connecting with multiple organs as it does so. *Vagus* means "wanderer" in Latin, which is fitting since it wanders around your body.

You actually have two vagus nerves, one on each side of your body, but these are referred to as a single nerve. Your vagus nerve contains around 75 per cent of your parasympathetic nervous system nerve fibres and controls your digestion, immune function and heart rate. It is also a cranial nerve that relays sensory information as sights and sounds and has a motor function – moving and controlling certain muscles.

The vagus nerve is critical in signalling to your brain that you are safe and all is well. As well as controlling your heart and breath rate, it can switch off feelings of stress and anxiety. The more vagus nerve stimulation you are exposed to, the higher your vagal tone becomes. This is important because your vagal tone is directly related to how resilient you are and how you handle and respond to stress. A high vagal tone also helps you to regulate your emotions, increasing both your empathy and compassion, as well as being beneficial for overall good health.

Unfortunately, with age and certain experiences, your vagal tone can decrease. Stimulating the vagus nerve is recommended for people with depression, and research is under way to see how it could help those with other conditions such as Alzheimer's disease[20] and rheumatoid arthritis.[21]

Sound healing is a useful, low-risk way to stimulate your vagus nerve. It also increases your vagal tone because of how close this nerve passes by your ears. As soothing vibrations from sound healing instruments enter your ear canal for your brain to decipher, the vibrations stimulate your vagus nerve. This nerve then sends vibrations to your internal organs, inhibiting the release of cortisol and adrenaline and calming your entire body.

Because your vagus nerve runs through the muscles in the back of your throat and is connected to your vocal chords, another great way to stimulate it and calm your autonomic nervous system is through humming. Once again, the vibrations are felt through this nerve and your parasympathetic nervous system is activated. Humming also encourages the release of the "love hormone" oxytocin and your body's painkiller endorphins and increases the amount of miracle molecule nitric oxide in your system (more about this on page 158).

SOUND HEALING AND RESONANCE

Sound has so many wonderful properties, as we will continue to explore across all these pages, but increasing the amount of resonance in your life is the golden ticket, a fundamental concept for this modality and the key to everything. Resonance changes your perceptions: how you relate to the world and how the world in turn relates back to you. It is about reaching your inner wisdom and living your truest life, connected to your highest self, and enables you to see the full potential in everything.

Stress and anxiety warp your perception of the world, making neutral situations seem dangerous and tricking your brain into thinking things are *not* good and that they will *never* get better. This can eat away and deplete everything that brings you purpose, joy and peace, dysregulating your nervous system.

When therapeutic sounds begin to regulate you once more, you become more clear-headed and have less space for irrational thinking patterns that would otherwise control and consume you. You have more space for resonance and begin to see things as they could be, at their highest potential.

As American author Wayne Dyer said, "If you change the way you look at things, the things you look at change."[22] This is resonance in action.

In physics, resonance is synchronization, when the vibrations of one thing match up with the vibrations of another. This is how all meditative sound works – your brain syncs to the therapeutic-grade instrument being used near or on you.

Resonance can be seen among people, too – when we say we "clicked" with someone and they are on the same "wavelength" as us. We resonate with people when we understand them and they understand us.

In sound healing and sound therapy, resonance is an "expanded, uplifted and positive state," according to Lyz Cooper, founder of the British Academy of Sound Therapy. This is the definition I am exploring here and the final context that is important for us. Resonance is a thought pattern: a recognition of the good. It is about how once you start seeing the best in the world around you, you notice even more of it.

Resonance is the basis of the popular social media concept "Lucky Girl Syndrome" (the idea that if you believe you are lucky, you will be lucky), which can be explained by neuroscience. Your reticular activating system (RAS) is a bundle of nerves in your brain that filters out unnecessary information and enables the important stuff – or what you determine is important – to get through. This is probably why fathers can sleep through their baby crying but will wake up immediately if someone tries to open the front door, or why the papercut does not hurt until you notice it visually.

Your RAS acts as a filter to the world and how you perceive it. It controls what you see. The more positive your filter, the better the outcomes in your life. This system enables the Law of Attraction and is why if you believe in the "angel number" 11:11, the more you see it and the more you will continue to see it. If your RAS thinks you are lucky, your brain will seek out information to confirm this – *making* you lucky. Your brain just wants to prove you are right, to seek out patterns to confirm your beliefs.

What being in a state of resonance does is improve those beliefs so that you become more optimistic. The safer your body feels, the safer your mind perceives the world; the safer your mind is, the more you have the capacity to recognize the good in your life. The more good you recognize, the more your brain seeks to find evidence of this – you expect positive things to happen to you, which invites better things to happen to you.

For this reason, sound healing is an effective way to reframe your thoughts and to switch up the narrative in your head and how you see things and diminish self-limiting beliefs.

Your "I can't do this" thoughts are no longer supported by your brain since the evidence of your life says "I *can* do this". As writer Anaïs Nin once said, "We don't see things as they are, we see them as we are."[23] Just because the obstacles are there does not mean your focus should be on them.

Energy is finite – you only have a certain amount of it – and like the adage says, "Your energy flows, where your attention goes." If you spend all your mental energy focusing on what is going wrong, you do not leave room to see what is going right. If you focus on the roadblocks, you narrow your vision and stop yourself from seeing the solutions. It is why, as optimism coach Simon Sinek explains, you tell skiers to follow the path, not to avoid the trees, because telling them that would only draw their focus to the trees, which they would then eventually hit.

Annoying, upsetting and irritating things will always happen. Embracing a more resonant-filled mindset is about choosing to see those obstacles that can clutter up your mind in a different way. This mental reframing also enables you to tap into the manifestation mindset. If you want to manifest a goal into your life, you need to first change how you think.

The Essence of Resonance

A few years ago, I was thrown into chaos when my neighbour flooded my flat and refused to get it fixed, give my plumbers access to his home or contact his insurers to get things sorted. It was an incredibly stressful time. I had to move out, my home was badly damaged, calls to lawyers consumed my days and bills began to stack up. I was so stressed out on the day I moved out, I was physically sick on the street. To say this event dysregulated my nervous system would be putting it mildly. In the end, when it was eventually fixed, I needed two walls replaced and I was seriously out of pocket.

Yet when I look back at that time, I never think about these things or how the situation came about. In fact, I have had to consciously sit here and recall these events. Instead, I just remember the moments of joy that this disaster afforded me.

I think about how lucky I was to be able to live with my nephew for four months. I am thankful for all the cuddles, the morning giggles, all the bottles I was able to feed him while looking after him, the forehead kisses and our countless walks to the park – all those tiny memories we made together that would never, under any normal circumstances, have happened. I also think about how much my nice neighbour Amie helped me: the phone calls she made and the car trips she took. I can truly say these are the moments I choose to hold on to. This is the essence of resonance.

Of course, I do not deny an element of privilege was at play in my situation. I have a sister who had both the space and desire to help me and take me in. I do not want to brush over that at all. But even still, I could focus on how I lived out of a suitcase or wore the same clothes for months, or how many hours I chased different people to get things fixed. I could choose to *only* remember the negative things that happened and not focus on the unexpected joy I experienced.

If you are wondering whether this means I forgave the neighbour, absolutely not. This is not a case of holding a grudge or being "low vibe". Seeing more resonance in your life does not mean you become blind to situations where you have been wronged. You can still recognize if someone has behaved appallingly. You do not suddenly become a beacon of love and light where nothing bothers you just because you attend sound baths, and you do not become a Buddhist monk or Mufti Menk because someone played a gong nearby. You continue to have human experiences – anger included. It just means that you focus on the good things that happen; you do not dwell and let yourself be consumed by the bad when that happens instead.

BECOMING MORE RESILIENT

I do not want you to read these pages and think that sound baths will stop you from experiencing bad things; they won't – that is impossible. Situations that you do not like, find uncomfortable or make you unhappy will always happen. This is just a part of life.

Having more resonance in your life also does not mean that you cannot recognise and address negative things when they happen, either. Thinking in that way will soon see you fall into the realm of toxic positivity. Life can't always be #goodvibes, and thinking that it will only set you up for failure and disappointment.

By increasing how much resonance you see in your life, sound healing decreases your reaction to resistance. You become so busy and grateful noticing, celebrating and revelling in the good that the bad is merely a redirection, a bump in the road. I am not talking about things that deserve the full weight of your feelings, such as bereavements (though we will discuss this further later on), loss and hurt, but I mean the small annoyances will just not matter as much. If someone pushes past you in a coffee shop, you will not care, or if a work project needs to be tweaked – okay, no problem.

Even when situations are challenging, they will appear to be more manageable and you will be primed to see a way out of what is troubling you, because how you react is what counts – and this is important to remember. When your nervous system is regulated, you are more likely to keep calm, which allows you to react better and from a place of balance. You are not on the edge all the time. This in turn builds your resilience and allows more magic to happen. Life's downs can't be avoided, but recovering quickly and getting up from the knocks will help you live a happier, more fulfilled life.

Often people mistake resilience as an ability to endure a great amount of stress, but this is not accurate. Resilience is how quickly you can return to a happy baseline and bounce back *after* stress. It is not how much stress you can put up with or how much negativity – be that from people, experiences or situations – you will tolerate. Building up your resilience will mean that you are not disappointed or even floored when faced with the unexpected.

IDENTIFYING AND RELEASING TRAUMA

Earlier I mentioned that one of the benefits of sound healing is that it affords you perfect privacy. Privacy is a good thing in a therapeutic setting because it enables you to be honest with yourself. Sometimes you may self-censor, even if you do not mean to. You may not even realize that you are doing it, as self-censoring may be so ingrained in your life and you are used to just soldiering through.

This type of self-censoring behaviour also means that you may have pushed down the real reasons for your feelings of stress or anxiety. In a sound bath, you access a subconscious part of your brain where societal and cultural expectations and your own blocks are stripped away. You can give yourself space to delve into the deeper parts of your consciousness, and this is where you can uncover deeper issues, triggers and traumas.

Until you know what is happening, what is stuck in your body, you can't fix it. You can't give it resolution because you do not know it is there.

Sound healing reminds you that it is okay to feel your emotions. You can sit with them, allow them to surface and understand that, like waves, your emotions rise and fall with intensity. It is important to remember that your emotions are full of information, too. They contain signals, clues and signposts, so truly allowing yourself to feel them is not a bad thing. No emotion is negative, as such; it is the suppression of them that causes the problems. When you hide from your emotions, push them back and avoid addressing how you are really feeling, you can't change the situation that has caused them in the first place. Feelings of trauma can have the same impact. Sometimes you may experience something that shakes you to your core, so you may have silenced it to ensure you can cope. The safe space of a sound bath can allow you to admit, or even uncover, your true feelings.

EMBRACING YOUR THOUGHTS

Before every sound bath, I tell my clients that if any thoughts arise in the session, this is okay. A common misconception is that sound healing is the same as other types of meditation, when a practitioner may expect you to clear your mind. This is not true and it is important to remember that thoughts, like emotions, are not inherently bad or indeed good.

Removing the pressure to keep a clear mind serves many purposes, especially making it easier for you to keep your mind clear. When you tell your brain to clear itself of thoughts . . . what happens? You are flooded by them.

Author and inspirational speaker Simon Sinek demonstrates this effectively in his talks.[24] He asks the audience to not think of an elephant. What inevitably happens is that everyone, including you reading these words right now, thinks of an elephant. This is similar to the way we now understand that we should tell toddlers to "walk, walk" instead of telling them to "stop running". Our brains simply can't comprehend the negative. This inability is called "The

White Bear Principle/Ironic Process Theory". If you go into a sound bath determined not to think of anything, you will think of everything. Remove that pressure from yourself – I give you permission.

Plus, your thoughts may lead you to answers. In sound healing, we believe that the answers lie within you and that your innate wisdom will guide you to the resolutions you need. Somewhere along the way we have lost the ability to listen.

If thoughts pop into your head, go with them. Sometimes this may be the key to taking you into an ASC, too. If your brain keeps bringing you back to something, the solution you need could be just beyond those thoughts, so do not block anything.

It may feel weird at first if you are lying down and listening to the instruments being played and you are thinking about the yoghurt you left in the work fridge. You are at a sound bath to switch off. You do not want to think about work, let alone something completely random. But go with it – let your thoughts come, view them with compassion and allow them to flow back out of your mind with ease.

When you are in a flow state where thoughts can come and go freely and easily without judgement, something incredible happens: other thoughts will flow in. This may seem random at first and may happen in your first or second sound bath (depending on how stressed or anxious you were when you came in), but think of it as free journaling happening in real time in your brain. (For more information on journaling, see page 228.)

One seemingly obscure thought may lead to another and then another and then all of a sudden you may come to something that does make sense. Suddenly the metaphorical dots may join and the real cause of your anxiety may surface. If Janice from Accounts comes into your head, for example, and you allow your thoughts to go with it, it could be that your recent general feeling of "meh" or stress actually harks back to an interaction you had with her but you did not recognize the link at the time.

COPING WITH DISCOMFORT AND UNEASE

Sometimes a sound bath will not be relaxing – you will still access many of the benefits of rest, but rest and relaxation do not always go hand in hand with this type of healing. Healing is not always rose petals and Insta-worthy tears that catch the sunlight as they fall from your face. Instead, it can involve ugly crying and feel challenging. At times, healing is uncomfortable – it is important not to shy away from that and to understand that feeling like this is informative and sometimes necessary on the road to reducing stress and anxiety.

Whether you are a seasoned sound bath attendee who has always reacted well to any sound meditation, or a total first timer, you should be aware that there will always be the possibility that you may feel uncomfortable or uneasy at some point in your sound healing journey, or even after a session.

Even if you are only using sound healing to access a place of rest and are not doing any trauma work alongside it, sometimes there can be blocks stopping you from returning to a place of stillness. Since a group sound healing session is designed to be restful, certain styles of playing will not be present – or will be heavily limited – in your session.

The focus is on harmony. However, I have also seen clients react less positively to the most harmonious intervals: the mix of notes – the really relaxing ones, the universally pleasing ones (the perfect fifth) – triggers a reaction that leads them to a place of "erm, not for me", taking them out of a trance and leaving them feeling unsure.

So why does this happen and, more importantly, what could it mean?

In sound therapy, we would call this Ego Dissolution or Threat of Ego Death. Before any panic sets in, do not be afraid because the word *death* has appeared. Just like a tarot pull, death does not mean the end here. Also, *ego* in this context is not about how self-important you think you may be or that you have a grandiose, inflated opinion of yourself. When we speak about ego here, I am referring to your conscious awareness, the thinking part of your inner self. This is the part of you that judges your surroundings, problem-solves and processes information – your belief system.

Your ego is there to protect you and keep you safe. If it snaps you out of a meditative state, this is usually because your ego is perceiving deep relaxation as a potential threat. But how could resting be a threat?

Imagine you are spinning plates like a circus performer. These are expensive porcelain plates and some may even have other crockery or towering amounts of food on them. For many of us, this is how our daily lives seem; we are on a continuous loop of spinning plates. Your arms may be deeply fatigued, which is why you have come to a sound bath in the first place. Instinctively, you know you need to rest but somewhere deep inside you are worried about what will happen if you stop spinning the plates – that they may all fall and smash, or if you put some plates down you will never be able to pick them up again.

This is when your ego steps in. It does not want you to smash the plates or not be able to pick them up again. It is trying to keep everything in your life going and it does not want you to fail. So what does your ego do instead? It snaps you out of your meditation, wakes you up and says, "No, thank you, this meditation business isn't for me." It makes you feel uncomfortable.

Alternatively, your nervous system steps in and wakes you up. Sometimes, especially if you suffer from chronic stress and anxiety, this becomes your baseline – a familiar feeling. There is safety in the familiar and any movement away from that can feel scary.

Stillness can feel unsafe, and surrender can feel terrifying. But if you give in to this, deciding after one experience that sound is not for you, the familiarity of stress will keep you stuck and hold you back. It may even stop you from achieving your goals and embracing the calm that you seek.

None of this is a problem, of course. If anything, this response shows how beneficial rest will be for you and how much you need it – to be able to put some of the plates down. If this happens in class, speak to your practitioner. Allow them to reassure you, giving some of the responsibility in that moment over to another person, and come back to try again. Let yourself ease back into it to prove to your inner self that everything will not fall apart if you *allow* yourself moments of pause.

As for your nervous system, the more you expose it to moments of rest, the more your body will come to learn that rest can feel good. By effectively rehabilitating your nervous system, you can show it that slowing down is safe and stillness is pleasurable.

Sound Bite

Dissonant intervals

Further up, I have mentioned how the perfect fifth – known as the most harmonious interval – can elicit a less than favourable response in a tiny number of people. So now let me share one of my favourite stories about dissonant intervals. These are the intervals that are supposed to cause a sense of unease and suspense and can seem harsh and cold. The most dissonant interval is the tritone.

Legend has it that the tritone sound was thought to be so uncomfortable that the Catholic church believed that only the devil could have come up with it and its use was banned in church music. Now, there is some debate about whether this is just a musical myth. But as someone who purposely avoids playing a tritone in my sessions, I like to think it is true!

ACCESSING DEEPER EMOTIONS

I am going to share two examples with you here about how thoughts may lead to understanding the cause of your stress, anxiety or trauma.

One client came to see me feeling like she needed a pick-me-up. As a mother with a young baby at home, she knew she was tired and adjusting to her new life and responsibilities. We had a chat after I had chimed the last of my instruments and let the gravely tones of my numerous shakers lull her back into an awakened state.

Six months prior to our session, she had been in a traffic accident, where someone had slammed into the side of the family vehicle. The baby, only two months old at the time, had been in the car and fortunately everyone was fine physically. The other driver had been at fault and the insurance issues were eventually resolved. She thought she was over the incident (which had been incredibly traumatic), yet this memory came up as she lay under the soothing safety blanket of Himalayan and crystal singing bowls. Like a garden with neatly cut grass, the evidence of the dandelion was not there for the naked eye to see, but the roots were still firmly in place.

This client was confused that, so many months on, with everything sorted, this traffic accident came into her mind. Physically she and her family were all fine, but mentally and emotionally deep inside she was still holding on to this experience. This is an example of how time does not matter when it comes to addressing stress, anxiety or trauma. Even something that happened six months, six years or six decades ago can leave a mark somewhere deep inside.

In his 2014 book, *The Body Keeps the Score*, trauma expert Bessel van der Kolk discusses in detail how the body can remember trauma long after our logical minds have insisted we are "over it".[25] This concept is not necessarily new, but it is often just ignored, forgotten and glossed over.

This brings me to the second client. She was stressed and anxious and everything was overwhelming her. The three weeks prior to her breakthrough session, her complaints of how she was feeling were nearly the same each time: anxious, overwhelmed, stressed out, triggered by everything and nothing in particular. By week four, she completely surrendered to the sonic work we were doing and the clarity came.

With her permission, I am going to share a small snapshot of what she experienced in that session, what her ASC state showed her and what she described to me afterwards. This was many, many years ago, but what she described was so profound that it has stayed with me with laser-sharp recall.

She described her anxiety as dark moths flying in her stomach: everything was dark, black in colour. As she moved metaphorically through the session, she found that the moths became bats that were flying from side to side. Then, slowly, between the heart-opening intervals and the expansive depths of the crystal singing bowls, everything began to turn midnight blue. As the sounds crescendoed, the bats in her mind's eye flew out of her stomach. She was enveloped within a green and yellow aura, likening it to the Northern Lights. Everything came in waves and swayed and swirled around her. Finally, she saw white butterflies, not flying but gently floating in and settling, turning everything white. The source of her anxiety – what had really been bothering her – became clear.

This is another perfect example of how you can feel as though you are stressed about work, your partner, studying, *insert your own example*, but as soon as you strip everything back and peel away the layers, the truth of what has really been causing your struggles comes to light.

If either of those clients had suppressed a thought – an act in itself that took them into the thinking, concentrating, focused state – these revelations would not have happened. New thinking regarding trauma already shows us that we need to flip the way we see healing, that we need a body-first, bottom-up approach. We can't just think our way out of stress and anxiety. Because sometimes, as these examples show, we consciously do not know what is causing that deeper layer of anxiety.

Sound enables you to access that.

PROCESSING LOSS AND GRIEF

Grief can feel so big, so all-consuming, and it is often difficult to find or know the words to describe it. You may not even want to say the words out loud. This feeling can be even more compounded if the bereavement you are experiencing – such as a miscarriage – leaves a hormonal imprint on you.

Loss can also present itself in other ways. The breakdown of a long-term relationship or marriage, the loss of a job, or losing one's home are all traumatic events. When you are in the middle of one of these situations, you may not even know where to begin if you try to describe it.

Once again, sound, through regulating your nervous system, becomes the perfect container to enable you to process such things without having to use your words. You can "just be" and indulge in a state of suspension where all worldly hurt starts to fade just a bit more each time.

Clients have described how insurmountable their loss felt before a session but then was slightly more manageable afterwards. One particular client phrased it in the most beautiful way. She spoke to me after a crystal singing bowl sound bath – her first time in a sound meditation. As I played the bowls, she said she felt as though she "breathed in joy and breathed out heartache".

You will not get past any type of heartache in one session – and nor should you expect to. Grief is meant to be felt and will take time to process. However, sound will help to chip away at the gigantic iceberg in front of you, and although some grief never really leaves, with time you will be able to sail past it.

FINDING CLARITY, CREATIVITY AND FOCUS

I previously mentioned that I wrote my business plan after a drum circle. It was my first-ever drum circle so I did not know fully what to expect. The session did not disappoint and, as the reverberations swirled around the room and permeated into my cells, it was as if I was being sonically shaken into action.

The drums were like axes chopping down everything that was camouflaging my vision. My mind was full of ideas and I could see what I wanted to achieve and how I was going to get it. When I "came round", I stayed silent and quietly grabbed my notebook. I poured all my thoughts into it before I had a chance to forget anything.

When we are stressed and anxious, survival becomes our main goal, our top priority, and there is no room for creativity or clarity. The body thinks it is in the middle of a fight, that it is in danger. You can't create from that place of being; you need to experience stillness for clarity to come. Stillness is an invitation for your thoughts to organize themselves.

The more you get caught up in that fight-or-flight state and let the daily humdrum of life take over, the easier it is to lose focus. That focus could be related to your work, goals, what you want to manifest or simply what your true desires are for the type of life you want to lead.

Your judgement can become cloudy and you become unsure of yourself, the uncertainty hammering home any self-limiting beliefs you have – the "I don't know how" quickly becoming "I can't do it", before turning into "I'll never be able to do it". This is when you may begin to lose yourself.

Look back at any times in your life when you were anxious – this may be right now. Notice how everything merged into one. Is it hard to remember what life was like before you were gripped by these feelings?

Therapeutic sound is a brilliant way to cut through the noise and shed everything that is holding you back, and this is all to do with being in a theta brainwave state. Theta brainwaves are slower than the relaxed alpha waves and come just before deep sleep hits. In essence, they straddle the tiny space between consciousness and unconsciousness and are the brainwaves of light sleep and dreaming.

Being in a theta-dominant brainwave state enables you to access your creativity. This is also essential for information processing, which means that when you are in a theta-dominant state, your brain has time to organize your thoughts. It is also the place of deep mental rest.

You also naturally experience this state and the corresponding bursts of creativity that can come with it. Have you ever had a brilliant idea just before you fell asleep? This is the same. When you are in a sound bath, you can maintain being in this state for longer periods than usual, therefore amplifying the benefits. Shamanic drumming produces the same frequencies as theta waves (3–8Hz). Combine this with how pleasing your ears find the rhythm, and it is no wonder that your brain wants to sync up quickly to this slower frequency.

Even as a practitioner I am not immune to this. In fact, out of all my instruments, my low-toned Bahia drum is probably the most challenging when it comes to not putting myself into a trance at the same time. Binaural beats, where a new tone is created from the difference of two near-similar frequencies, is another excellent way to induce this state, although you can also get there with gongs and Himalayan bowls.

Theta waves are key for good brain health. Your brain needs the mineral potassium to function optimally, to help your brain communicate with cells inside it and around your body. Your potassium levels are uniquely in balance with your brain's sodium levels. Being in a beta brainwave state for too long (such as when you are concentrating or feeling stressed and anxious) disrupts this delicate balance and leaves you feeling mentally drained and foggy. However, when you are in a theta-dominant state, your cells reset your brain's correct potassium to sodium ratio, leaving you feeling clear-headed and mentally refreshed.

BOOSTING YOUR IMMUNITY

Stress can affect your physical health, from headaches and migraines to skin rashes and hives, tummy issues and even a weakened immune system. Chronic stress triggers inflammation in your body, so reducing inflammation is crucial to protect against disease. Having a high vagal tone through vagus nerve stimulation is anti-inflammatory, meaning that engaging in sound work can help to keep you healthy.

Your immune system is also highly sensitive to the stress hormone cortisol. In short bursts, cortisol can help you get out of a dangerous situation, but if your levels are chronically high, not only will you get ill more often but you will remain ill for longer. High cortisol levels from stress may also lead to weak digestion, resulting in poor nutrition absorption, and this too can leave you prone to illness. The vagus nerve, with its direct link to your gut from your brain, can help to counteract this.[26] Less stress equals fewer digestive issues and therefore also aids your body in helping it repair itself.

PAIN REDUCTION

Sound healing can help with pain management, reducing your relationship with pain in several ways. A drum massage, where a frame drum is played very close to your body, interrupts the pain signals from your body to your brain. Even though the drum does not touch you, you can feel the vibrations travelling up and through your body, and this is deeply relaxing. Drumming also promotes the production of endorphins, your body's own painkillers.

Even if you are not having a vibroacoustic massage, you can still experience reduced pain. Preliminary studies have shown that attending a sound bath reduces participants' perception of pain.[27] In many ways it is the knowledge that sound and music can be used for pain management that brought me to sound as a healing agent years ago (more about this in Chapter 5).

BUILDING CONNECTION

Of all the benefits, building connection does not get talked about enough, and sometimes it is not even acknowledged at all. Community and having a sense of connection with those around you is a huge healing component of attending sound baths. "Attending" is essential here – this can't be done from the comfort of your own home using a good pair of quality headphones.

The more sound baths you attend, the more your reality will start to shift. Also, your viewpoint is likely to become more positive for all the reasons we have discussed and you will be less stressed as a person. When this all happens, clients sometimes find there is another more subtle shift in their personal lives, too – a change in their friendship dynamics. The more you increase your ability to let go of things that no longer serve you and see the resonance in your day, the more you will notice how much you have not only changed but how different you may have become from certain friends.

Do not get me wrong; you will not go to a few sound baths and then lose or even ditch all your long-term mates. But I firmly belief that your vibe attracts your tribe. Your vibration

– the thoughts you put out into the world – will attract the people you surround yourself with. This means that when your thoughts start to change, you may find that you want your company to change a little, too.

We all know that misery loves company, and there is nothing like off-loading and whinging about something to those around you. In recent years, research has proven that feelings can be contagious.[28] Stress is also contagious. When you are chronically stressed, there is a strong possibility that so are most of the people around you. This is a psychological effect known as "emotional contagion".

Similar to how women sync up with their periods, you may sync up with those around you in other ways. Science shows that cortisol (the stress hormone) literally seeps out of your skin into the air, so the people you spend the most time with – both male and female – absorb it into their systems. The evolutionary reason for this makes sense: if there is an incoming threat, it is best that everyone in the tribe knows. If you can all react quickly and together, this ensures a better chance of survival.

When you start your healing journey back to your healthiest self and begin to regulate your nervous system, you will naturally shift towards becoming a more positive person. Your outlook will no longer be clouded by as much resistance and the resonance in your life will dial up. It is also a case of physics – be it the Law of Attraction or the Law of Assumption, more positive life experiences will pour into your world. Inevitably, you may find that the person who you had things in common with before no longer fits in your new life. Your vibe changed, so your tribe may have to do some shifting as well. This is not a dramatic shift that happens with a huge argument. It can be as simple as being at dinner one night and realizing that you simply do not have anything in common anymore with a particular pal. Or speaking on the phone and finding yourself slowly uninterested. Think of this as drifting away on still waters rather than a turbulent storm.

I want to interject here and say that moving away from some people, even old friends, is not a bad or scary thing. This can be part of a positive journey so embrace it. Do not see this as having to cut people off; see it as an opportunity to meet new people and make new friends with whom you may embark on a brighter adventure.

It is no surprise that so many clients who come to group sessions end up becoming friends. You see these people weekly, you share this almost sacred space with them and you revel in the vibrations of these healing tools. This is why I always tell clients to speak to their mat neighbour after the session. If you are seeing the same people attend a sound bath, or even if someone is sitting by themselves and you are seeing them for the first time, go and say "hello".

You would be surprised that even in a city such as London, which is probably not unfairly stereotyped as being unfriendly, how easy it is to strike up a conversation after a sound bath. If you have gone with friends, you can grow together, beat stress together and heal together – but keep your eyes and hearts open to making new friends, too. There is no easier pool to find friends in than the one that is causing you to positively grow. Why? Because the people there are doing the same thing. Whether you see it as building a new community or simply extending your current circle, your fellow sound bathers are the perfect crop of people to do it from.

Connecting with other people boosts your immunity too. We are social animals, supposed to move in packs, and isolation and living alone has been shown to be as bad for our health as smoking.[29] The quality of our non-romantic relationships, of our connections, are proven to help protect us. Inflammation is the buzzword du jour when it comes to health and we need to be wary of it, as it is the driver of disease. Did you know that lacking strong, meaningful social connections can actually drive up your chances of increased (unwanted or needed) inflammation?[30]

As a natural extrovert, I know it is easy for me to say hello to a stranger, but if you are on the introverted side of the coin, know that you have already done the hardest part – you have already conquered any awkwardness or shyness or discomfort by attending a group sound healing. Most sound bath spaces have areas that you can stay in afterwards for a glass of water or some herbal tea, so do not dash off in a hurry after a class. Stick around and embrace the possibility of meeting a new friend.

Having a regular in-person practice does not just open the door to possible new friendships; there is another highly beneficial reason to do so. The more sessions you attend, the deeper your practice becomes. Your brain knows what to expect, so with each session you can explore more of yourself in that altered dimension of your brain. Regular sessions also give the practitioner a chance to get to know you energetically so they can personalize your experience – even in a group setting.

IMPROVING YOUR MOOD

With so many vast and varied positives that sound healing provides, I nearly forgot to mention the most obvious one, the one that does not involve bells and whistles. This is also the simplest, most beneficial positive of sound healing – it improves your mood. Sound allows you to stop and pause, its ripples encourage rest and the sonic vibrations invite ease. When this happens, you feel profoundly relaxed.

Improved mood may be a sign of ASC, but it can happen even without it. In fact, the most common thing people say when leaving a sound bath is how much better they feel, how much more at peace they are after a session and how much happier they are than when they walked in. If sound healing did nothing else but this, it would still be worth it.

CHAPTER FOUR

"I go to nature to be soothed and healed, and to have my senses put in order."

– John Burroughs

The Sounds of Nature

What are your favourite memories of being on holiday? Are you outside? Are your feet in the water or beside the ocean with the lapping sounds of waves? Maybe you are on a hike outside, among dense trees.

For most of us, the moment we get the chance to step away from our desks and our jobs, we head outside. Humans are hardwired to want to be in nature. While our modern lives may be very different to how our ancestors lived, our biology is still the same. Being in nature is deeply restorative, with multiple benefits for our health and wellbeing.[31 32]

Improved cognition, a decrease in anxiety and depression, and better heart health may be the more widely known benefits, but did you know that spending time in the natural world also makes us more altruistic and kinder to both each other and the environment?[33]

One study by the University of Derby in England also found that those who feel connected to nature have a greater eudaimonic sense of happiness, which is when your life has a sense of meaning and purpose.[34] Another study in the *International Journal of Wellbeing* concluded that "connecting with nature is one path to flourishing in life".[35]

Nature is so good for us that cultures such as the Japanese have a whole wellbeing concept around it: *shinrin-yoku*, also known as forest bathing. In Scandinavian countries, children's first experience of education is not in a classroom but at Forest School. They spend their days playing, learning and running around outside among the trees. This may sound like a foreign concept to many of us living in Western countries, but this is exactly how our ancestors would have spent their early, formative years.

These examples may be cultures apart, but the idea is that spending time surrounded by trees and allowing your mind to wander as you sit or walk (and play) among the greenery is good for you. So why exactly do you feel better in places with dense greenery?

The rainbow of greens and browns relaxes your eyes while the presence of phytoncides – essential oils given off by trees – strengthens your immune system.[36] These natural chemicals keep the plant free from bacteria, insects and fungi and also boost the activity of natural killer (NK) cells in human bodies. NK cells seek out and destroy cancer cells and viruses.[37]

It is not only the sights and aromas of forests that can trigger your shoulders to drop and a sense of calm to wash over you; it is also the sounds. From the rustling of leaves, the gentle swooshing of the winds to the trickling of water in a babbling brook or the sweet song of chirping birds, Mother Nature herself has gifted us with an orchestra of sounds that our brains find both pleasing and soothing.

Sound healing is all around us in the natural world. When we retreat into nature, away from the humdrum of manmade noise, we get the opportunity to connect with it. Not only does this natural orchestra sound good to us, but it feels good to us, too. It is restorative and boosts our mood. The sound of the wind blowing through the trees even has its own word – psithurism. One study has found that after just ten minutes of listening to birds singing – live or recorded –

people reported an increase in their feeling of wellbeing.[38] Combining birdsong with the other gentle sounds of nature increased the positive effects. Researchers believe the reason we love birdsong so much is because it indicates a place of safety free from predators. If nature is thriving somewhere, this means it is also safe for us to thrive in that environment.

SEAS, TREES AND FEELING AT EASE

The forest is not the only part of nature that has its own healing sound. From the grounding and repetitive lapping of the ocean to the melodic splashing sounds of the sea, being near and around large bodies of water also helps us to destress. Hippocrates, the father of modern medicine, recognized this with his thalassotherapy (sea therapy) work.

The low frequency of these sounds signals to your brain that you are safe. Because you hear them for a sustained amount of time, brain entrainment happens. The rhythmic crashing of waves slows down your brainwaves to lull you into a state of deep relaxation. The human body is designed, and has evolved, to respond well to these non-threatening sounds and find them calming while signalling to the brain to reduce the production of stress hormones.

Spending time in nature is a brilliant way to calm your nervous system. This is becoming increasingly important since human-made, anthropogenic noise has the opposite effect. Noise, which is unwanted sound, is all around us in urban spaces and is impacting our health in a negative way. Sustained noise exposure, such as we find in cities, leads to an array of health conditions. Disturbed sleep and general annoyance may be the obvious symptoms, but studies have also shown that being exposed to constant noise can lead to an increase of high blood pressure and even cardiovascular disease.[39]

People recovering from stress in places with high levels of noise pollution also take longer to recover than those near the sounds of natural environments.[40] Scientists have even found that urban noises such as traffic can impair how well you perform in tests.[41] Finding time away from urban soundscapes is something we should all try and make time to do.[42] This may involve driving somewhere for a walk in a park or forest or by a lake.

Sound Bite

Thunder and Lightning

The sounds of nature are wonderful for changing brainwaves and bringing you to a place of ease. So it is likely (especially if you have had a session with me) that you will have heard elements of this recreated for you in a sound bath, from rain sticks that mimic the trickle of water to the rumble of an ocean drum that laps at the imaginary shores inside a treatment room. There is even an instrument to replicate thunder – aptly called the thunder tube. If you hear these soothing sounds during a sound bath, enjoy them. I often see clients opening their eyes to figure out what each sound was or even how I am making them. Let your thoughts go and let the sounds take you on your sonic adventure.

HOW TO LISTEN BETTER

Whether you are in a sound bath or out in nature, you can work on *how* you listen. Hearing and active listening are two different things. You can hear a TV in the background while you work, but you may have no idea what is being said because you are not listening.

Unless you have a medical condition, either genetic or damage that has been done to your ear drum or cilia (the tiny hairs in your inner ear), hearing is automatic. The vibrations from the source of the sound travel through the air to your ear drums. These vibrations are carried through your ear and translated by your brain, so you know a sound or several sounds have been made. But listening takes effort. It is a conscious activity that requires attention and demands concentration. Deep listening is a skill that can be worked on, and doing so will improve your experiences of spending time in nature.

While I do not suggest you *concentrate* in a sound bath, having the ability to actively listen will help you in a sound healing session. If an unexpected noise or interruption breaks you out of your meditative state, having the skills to listen will help you tune out the noise and return quickly to the therapeutic sounds being offered.

Active listening is also a mindfulness technique, bringing you back to the present. Improving your proficiency in active listening will enable you to tap into the sound healing all around you, too.

ACTIVELY TUNING IN

Your step-by-step guide to mindful listening:

1. Head to a woodland, forest, body of water or park, somewhere with nature.
2. Sit or even lie somewhere comfortably and relax your shoulders, arms and legs. If you are holding any tension in your face, let it ease and soften your body. Take in your surroundings for a few minutes as you do so, observing where you are and appreciating the visual beauty of your space.
3. Now close your eyes.
4. Bring your awareness to the sounds around you. What can you hear?
5. Let the sounds come to you, one by one, and focus on the sounds closest to you first. Are the leaves fluttering? Can you hear birds calling to each other? Is the river moving slowly or at speed? Observe.
6. Now draw your attention to sounds further away from you. Are there people nearby? Is there a plane in the sky?
7. Finally bring your attention to yourself. Can you hear yourself breathing?

You do not need to attach any feelings or thoughts to what you are hearing, to intellectualize the experience or to attach judgement to it. If there are people nearby speaking loudly or having a spirited conversation, this is fine. You are just observing.

Feelings of gratitude are the exception, as you can add this sentiment to every situation, sprinkling it on liberally and wrapping yourself up in it. If the chirping of the birds sounds beautiful to you, appreciate it. If you live by a river or lake, allow yourself feelings of gratitude that you are able to access spaces of nature and their harmonious sounds with such ease.

Embrace Your Environment

It may not always be possible to head off to somewhere with rolling hills and vast bodies of water, but that does not mean you can't practise active listening. You can do this anywhere, even in an inner-city apartment. Mindfully tuning in like this will bring you into the present moment. When you are stressed and anxious, you are either thinking of the past or the future. Mindfulness will ground you to the exact moment you are in, helping to reduce feelings of discomfort and making you present to both your immediate situation and surroundings.

Here is how to listen at home:

1. Sit comfortably and close your eyes.
2. Bring your awareness to your surroundings and listen.
3. Can you hear any birds outside? Try to tune into the sounds they are making. Is there a hum of traffic outside? If so, what is it telling you? Are the cars moving fast or are they stuck in traffic? Can you hear the clock ticking or your fridge humming?
4. When you open your eyes, notice if there were any sounds that you were surprised about. Had you heard or noticed them before?

GROUNDING

A lot has changed since the time of our barefoot ancestors, and you may find that unless you are on a beach holiday you never touch the ground with your bare feet. But if you are indulging in one of earth's many nurturing soundscapes, amplify the benefits of being in nature by taking off your shoes.

Grounding your body to the earth has been proven to reduce stress levels, improve circulation and even decrease levels of inflammation and pain.[43] Acupressure points in your feet are stimulated when you are walking on the ground without shoes, providing stress relief. Walking barefoot has many other benefits, too.

The human body is made up of atoms. When you are stressed and anxious, you end up with a detrimental, health-damaging positive charge. Coming into direct contact with the negative ions on earth's surface brings you back into balance.[44] Like sound healing, this also improves your vagal tone and reduces the stress hormone cortisol in your body. Since you are outside listening to nature's symphony, this gives you a perfect opportunity to combine your sound healing practice with grounding.

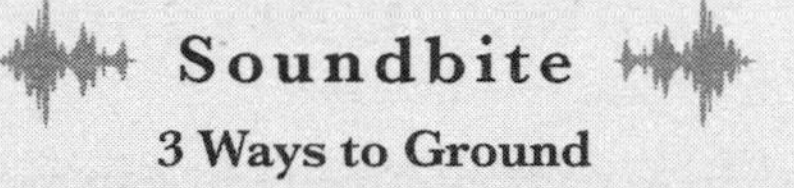

Soundbite

3 Ways to Ground

Grounding, or earthing as it is also known, is super easy. You just need to allow your skin to be in direct contact with the earth. Try and do this for at least 30 minutes.

- Walk barefoot on grass, moist soil, sand or gravel while you are listening to the natural sounds around you – whether this is birds or the rustling of leaves.

- Swim or paddle in a natural body of water such as the ocean or a lake, engaging all your senses. The water is negatively charged, so just being in it will ground you. While you are in there, notice how the water feels against your body and listen to the waves. Is the water lapping against the shoreline? Take it all in. The key is to feel restful and relaxed; this is not an Olympic swim meet, so only paddle, float or do gentle strokes in the water.

- Take up gardening. You need to be in direct contact with the earth so you will have to ditch the gloves. Gardening has an added benefit for your mood. When you disrupt the soil it stirs up its microbes, in particular *mycobacterium vaccae*, which stimulates the production of the happy hormone serotonin when inhaled. You can choose to tend to the flowers or veggies in your soil while actively listening to the gentle soundscape of your garden (or allotment) or simply let your mind wander aimlessly. Both are beneficial.

THE SONIC RAINBOW

As mentioned in Chapter 1, sounds can be categorized as different colours – all with their own signature of frequencies. Most examples of white noise are manmade, the result of technology. White noise contains all the frequencies that are audible to the human ear and this is not seen that often in nature. Because white noise contains the full spectrum of frequencies from 20 to 20,000Hz at equal strength, it is especially good for drowning out any unwanted noise.

Many of the sounds we have mentioned so far are examples of pink and brown noise – nature is full of these. Pink noise is softer than white noise and sounds deeper. This is because all the frequencies are present in pink noise but the higher frequencies are quieter. Examples of pink noise include rainfall, wind, the sound of rivers and rustling leaves; even parts of birdsong can be considered pink noise.

Pink noise helps you to fall asleep faster than white noise, gets you to deep sleep more quickly and improves the quality of your sleep, keeping you asleep for longer.[45] Nature's pink noises are soothing to your brain, which explains why it is so easy to want to fall asleep when you are among rustling trees. These sounds are pleasant and relaxing.

Pink noise has another surprising benefit. It can improve the memories of older adults who would otherwise be at risk of age-related memory decline.[46] This means that spending time in nature is crucial, whatever your age.

The deepest noise of all is brown noise. This is the sound of a heavy rainfall (or even a powerful, pounding shower), of crashing ocean waves, strong waterfalls and the low, rumbling of thunder. Brown noise is all about the bass: the lower frequencies stealing the spotlight and being the loudest; the higher frequencies getting quieter with each octave. Brown noise is great to switch off to, and some people say it also improves their focus.

The Colours of Noise

Colour	Example	Good for:
White	TV static, whirling fan	Masking unwanted noise, improving sleep, managing tinnitus
Pink	Rainfall, blowing wind, rustling leaves	Falling asleep faster, staying asleep longer and improving memory
Brown	Strong waterfalls, rumbling thunder	Deep relaxation, improved focus

These are the best coloured sounds for inducing a state of relaxation, but there are others, too. Sound can also be:

- Blue: A concentration of super high frequencies with a shrill quality.
- Violet: The opposite of brown noise, each octave getting louder.
- Grey: Contains all the frequencies, but the middle frequencies are not as loud, although the human ear perceives this to be the same as white noise.
- Black: The colour of silence.

THE RELAXING PURR OF CATS

If we were to settle the debate on whether cats or dogs were the better pets, purring firmly puts cats in the lead spot. Having a cat is good for you. It slashes your risk of having a heart attack or stroke and can even lower your blood pressure.[47] This is all to do with the sound and vibrations of a cat's purr; they are literally healing tools on four legs. A purr is between the lower frequencies of 20–100Hz, which we find restful, so having a cat is also good for calming your nerves and improving your mood.

Sound Bite

Nature's Alarms

Not all animal sounds have a soothing effect. The growling of a dog showing its teeth or the high-pitched squark of a seagull will certainly not induce any feelings of calm. Nature teaches us which sounds we should be wary of. The frequency of these sounds serves as a warning, prompting your brain to instruct your adrenal glands that you need adrenaline. As this hormone is released into your bloodstream, your heart rate increases, your pupils dilate and your body gets ready to react.

CHAPTER FIVE

"Without music, life would be a mistake."

– *Friedrich Nietzsche*

The Magic of Music

I will never forget the first time I heard Adele's "Someone Like You". It was 2011, and I was at home watching the BRIT Awards. I was back living in London, trying to decide if I was going to stay in the city or pack up my suitcase and go off on another adventure. There, under a spotlight, a piano began to play. There were no fancy backing dancers or visuals: just one instrument and one voice, *her* voice, introducing us to the song before building in both volume and musicality.

Sitting there, listening through the TV, that song touched my soul. I heard the lyrics and felt the music firing out of the speakers permeating directly into my body. Adele's voice raised the hairs on my arms, giving me goosebumps, and my skin began to tingle. I understood her sadness and felt the chords being played as my physical and emotional state changed.

I know I was not the only one to have such a strong experience, as "Someone Like You" continues to be one of the most downloaded songs of all time in the UK alone and that performance remains iconic. I am sure that you, too, will have had a reaction like this at some point to a particular song.

THE EMOTIONAL IMPACT OF MUSIC

Music is woven into the fabric of our days. It is always there and features in both our happiest and saddest moments. While Adele's lyrics also struck a chord (excuse the pun) with so many people, music without words can influence our emotional states and elicit a physical reaction as well.

Think of any horror film and you will notice how quickly the power can be stolen from a scene when the TV is on mute. How do you know a shark may be lurking in the deep waters and ready to attack, or that there is a knife-wielding masked psychopath behind the door? The music tells you. The choice of instruments used, the pitch and tone of a piece raise the hairs on the back of your neck and quicken your breath, leaving you with a taste of anticipation in your mouth. This indicates that something is about to happen and taps into your innate stress responses to prepare for it.

Music may not only scare you but it can also motivate you. Have you noticed a particular song makes you run a bit faster in the gym or peddle a little harder on the bike? Some music can help you on the path to healing heartbreak, allowing tears to flow when you connect deeply to an album, enabling space for release. Other songs, forever painted with the ink of love, become intolerable to a couple when they are no longer together.

A particular beat may inspire a movement, a war and embolden soldiers. It can also announce an arrival. Think about the music that a boxer may come out to for a fight. Now imagine how much of an impact they would have stepping into the ring if the *Teletubbies* song was playing – psychologically they would lose the fight before it began.

Music is there when you are feeling overjoyed and happy, too, adding or indeed creating those feelings. I love so many different types of music and believe you should be open to as much as possible as well. There is nothing wrong with listening to Dolly Parton one night and Snoop Dogg the next.

Like scent, songs can take us back to a particular point in our personal history. The Chordettes' "Lollipop" or Bill Haley's "Rock Around the Clock" will forever take me back to giggling with Bollywood-loving grandparents who knew all the words to those songs and sang them to me. In fact, certain old Bollywood songs are laced with cherished memories of a time of happiness, comfort and joy, when my personal world was brimming with so much unconditional love.

Nancy Ajram and Nelly Furtado remind me of the years I lived in the Middle East. Crowded House's "Don't Dream It's Over" will always take me back to being on a bus with newfound friends in New Zealand. The Topdeck tour guides did this on purpose. They played it every time we stepped on the bus and it was always the first song on the playlist. If that song comes on, while I could be anywhere in the world physically, mentally I am back on the coach whizzing past the snow-capped Southern Alps of Aotearoa. Meanwhile, certain recent songs are currently building new neural pathways in my brain to remind me in the future of all the fun that was had during the kitchen discos with my beloved niece and nephew.

I am sure many of these examples will resonate with you because even if the song choices are different, these experiences are a core part of life. We all know and understand these moments because we have lived them as well; music and our connection to it is inherently a human experience.

Music has a way to transcend time and space, making us both present but also whipping us away to a moment in the past. It transcends language, too. Music does not need lyrics nor does it require you to speak the same language for you to be moved by it. Even if you do not know anything about music, have never played an instrument or can't read sheet music, you can still take part in it. You can choose what you like even if you do not fully understand why you like it. You just know that you do. We are all drawn to music and in many ways we have all used music therapeutically in our lives – we just may not have identified it in this way.

THE PHYSICAL IMPACT OF MUSIC

Over the years, several studies have proven classical music to be effective in pain management. Cancer patients experience pain relief when listening to instrumental music.[48] Slow-tempo classical music has been found to ease discomfort from arthritic pain, according to research at the Florida Atlantic University College of Nursing,[49] and soothing music can even be used to recover more quickly from open heart surgery.[50]

Music's effect on pain relief is related to how your brain releases oxytocin when you listen to it, but other hormones are important as well. When you hear music you really like, your brain triggers the release of dopamine; this is why when you put on your favourite song, it immediately lifts your mood.

If your brain is wired a little differently[51] and you have more nerve fibres connecting your auditory cortex (the part of your brain that processes sound) to your anterior insular cortex (where your feelings are processed), you may also experience something called frisson. Frisson is a psychophysiological response that can occur when you listen to music. This is when you get goosebumps, the hairs on your arms and neck stand up, along with a tingling feeling.

Also known as an aesthetic chill or "skin orgasm", frisson happens because of musical expectancy and anticipation. Your brain likes to find patterns, so subconsciously when you listen to music you predict what will come next. These are micro-second predictions in your brain rather than something you are consciously doing. When your brain correctly predicts a change in pitch or tempo, you are rewarded with a shot of dopamine. The interesting thing is that if you are completely surprised by a change in the music in a way you did not expect, this can also be thrilling enough to warrant frisson.

Frisson activates your sympathetic nervous system, triggering a release of adrenaline as well – hence the goosebumps. The combination of dopamine and adrenaline makes this experience exciting rather than anxiety-inducing. In a sound bath, percussion instruments and gongs can be used to create frisson as the sound they produce builds in intensity.

A SOUND BATH OF ONE'S OWN

People regularly ask how I found sound, intrigued about my journey into this deeply restorative modality. They are also often curious to know if I have always been this relaxed.

My journey into sound is two-fold. Therapeutically speaking, it begins over 20 years ago, but perhaps, as mentioned above, it has been there from day one. My first conscious foray into sound, or rather music, as a healing tool happened when I was at university. I have always had a keen interest in health and wellbeing so I was constantly reading about these topics, even though I was studying English Literature. I stumbled on an article that explained how scientists had found that listening to certain types of classical music could help with pain management. *Wow*, I thought. *Music to manage pain – that's incredible.*

That summer, when I returned home during the holidays to have some wisdom teeth removed, I knew what my pain management plan was going to be. I was going to listen to classical music after surgery. So that is what I did. I played classical music on repeat. I recovered and returned to university – promptly forgetting that I had used music in this way.

It was only many years later through my work as a wellbeing journalist that I found myself in the sound healing arena once more. This time, I was invited to a group crystal singing bowl sound bath. I had no preconceived ideas before I went. It was crucial to my work that I always approached things with an open mind. I would have to form my opinion after the session anyway, and any questions I had would be answered then, too.

I lay down and the practitioner played the colourful bowls in front of her. I remember the room vividly but I can't remember her name or what she looked like. I also do not remember what happened in the session. My memories begin again at the end of the sound bath, when we were told the session had finished and we could get up when we wanted to. Everything was a blur. There was no way that session had lasted for an hour, I thought – I had just put my head down. The time had flown by. I knew I had not been asleep, but as I slowly gathered my things, I also knew that I was not fully "awake" yet either.

They asked if I had any questions.

"What? Who? Erm, I'll email you," I responded, and I left. I did not want to engage my thinking brain just yet.

Somehow in that hour I had been transported to another dimension. My internal monologue of things to do and places to be had stopped. I felt like I had slept, but it was not like an untimely short nap that leaves you groggy. I felt a deep sense of rest, akin to waking up after eight hours of sleep in fresh sheets on holiday. The powerful vibrations of the crystal singing bowls had evoked feelings of calm and relaxation that I had never experienced before. I went back, and then back again.

MUSIC VERSUS SOUND

Music can be described as the art of arranging sounds in a pleasing or emotive way and is a combination of tones organized with rhythm, harmony and melody. I am not a music therapist, as this is a completely different discipline to being a sound therapist. But while sound healing and music therapy are different, there is some crossover.

Music therapy uses music and a wide range of musical styles to express and support a client's needs – to process or even access emotions and feelings. A session may involve lying down and listening to music, singing, playing an instrument and even dancing to songs. The music created is improvised and you do not need to be a musician to attend.

A sound bath helps to process emotions too but is also about connecting to your body on a deeper level, unravelling what lies in your subconscious brain. Sound therapists play a series of sounds but we are never trying to create a melody, although in drumming sessions we use rhythm to entrain you. Sound baths also harness the power of vibrations to change your physical, emotional and mental state.

SOOTHE YOURSELF WITH SINGING

Perhaps the most important instrument on the planet is one we forget we always have with us – our voice. I have included singing here because it transcends sound healing and music therapy – it is both – and is an incredible tool for healing and happiness.

Your voice is powerful, whatever you choose to do with it – whether you are singing, chanting, humming or overtoning (a singing technique in which you manipulate the harmonic note you are singing). Like humming, singing also activates your parasympathetic nervous system. Your vagus nerve is connected to your vocal cords and the muscles in the back of your throat. When you sing, the vibrations from your voice directly stimulate this important nerve, which leaves you feeling calmer and less stressed.

Singing also impacts your hormones. When you sing, endorphins are released to make you feel good. Singing in a group or with a band takes this one step further. When we sing together, your body releases oxytocin, the love hormone that helps you bond and feel like you belong to a tribe or community. This is why choir singers frequently report feelings of joy and euphoria after sessions and why you will never see a gloomy face at a Pentecostal church – it is down to all the upbeat, group singing.

Our voices have healing powers. Ever wondered why you say "ow" when you hurt yourself? Studies show that vocalizing a sound that requires little effort (such as "ow") improves tolerance to pain, while keeping quiet feels more painful.[52]

Singing also benefits your body in a physical way. Belting out a tune makes you stand up straighter, improving your posture. As singing involves using many inner muscles, it tones your core and strengthens your diaphragm as well. Singing also requires taking deeper breaths, making it a fantastic aerobic activity that is good for your overall cardiovascular health. It therefore helps to increase your lung capacity. This is particularly important as in recent years having good lung capacity has been proven to be one of the best markers for longevity,[53] which means singing may even help you live longer.

Studies have also shown that singing supports your immune system by improving your levels of immunoglobulin A (IgA).[54] IgA is an antibody that helps to fight off the germs that make you ill.

USING YOUR VOICE

Even if you are not a natural singer, you can still reap the benefits of singing: sing out of tune, sing at home, sing in the shower . . . it really does not matter. If you are still not convinced that you will be able to sing, try humming, chanting and even gargling loudly with water instead. These all stimulate your vagus nerve and give many of the same benefits as singing.

I often recommend my clients hum between sessions, as this is a great way to introduce a sound healing exercise into their daily life with ease. Plus, you do not need any equipment to do it and it can be done anywhere.

Humming is beneficial to your health. As well as kicking your "rest and digest" system into action, it also produces an abundance of nitric oxide in your body. Among its many benefits, nitric oxide is responsible for expanding your blood vessels to increase your blood flow, while reducing the build-up of heart attack-causing plaque.[55] Having low levels of nitric oxide has been linked to a multitude of other ailments, too, including high blood pressure, heart disease, irritable bowel syndrome and even Alzheimer's disease.[56]

Sound Bite

How to Hum

To begin, suction your entire tongue to the roof of your mouth – your tongue should not be touching your front teeth but should be positioned a little bit back. Not only is this correct tongue posture, but when your tongue is resting up against your soft palate, it kickstarts the vagal tone process.

Keeping your lips gently closed, take a deep breath in through your nose. Exhaling out through your nose, make a firm "mmm" sound – your hum should continue for the length of your exhale.

Repeat the humming for a few minutes or until you feel a sense of presence and calmness sweep over you. You can be seated if you wish, but equally you can be standing in a lift, on the street or on a train. The beauty of humming is that it can be done anywhere.

Sound Bite

Chanting

When you think of chanting, which is a mixture of both music therapy and meditation, your mind may turn to Buddhism or Hinduism. But chanting features in all world religions including the Abrahamic ones such as Judaism and Islam (in which it is called dhikr). Ancient civilizations have also incorporated chanting into their rituals and daily lives across millennia, from the ancient Egyptians to the indigenous Australians, and for good reason.

Repeatedly saying a phrase or short sentence, usually a prayer or mantra, is the basis of chanting, and countless studies have extolled its virtues. Chanting helps to reduce feelings of stress while improving your mood.[57] It increases the blood flow to the brain[58] to ward off degenerative brain diseases and it can even make you more selfless and compassionate.[59] In essence, chanting helps to connect you both to yourself and those around you.

Chanting is primarily a religious experience and the best results come when the person doing the chanting truly believes the phrase they are saying. If you have faith, then choose what means the most to you. If not, then you can choose to repeat "aum" ("om"), which is taken from Hinduism. Although Hinduism is not a closed practice, you should be respectful when using "aum", and correct pronunciation is vital. The power of "aum" lies in its vibrational quality. Its frequency is said to be a healing 432Hz, but this is only true if you are saying it properly.

FIND A SINGING SOUND THERAPIST

I do not have a good singing voice, so including it in a session would not be calming for anyone involved – my voice firmly belongs in the shower. But if you find a sound healing practitioner who does use their voice, this can be an incredibly intimate and moving experience. Our brains respond really well when we are sung to,[60] so it is not surprising that parents instinctively sing lullabies to their babies.

AUTONOMOUS SENSORY MERIDIAN RESPONSE (ASMR)

Autonomous Sensory Meridian Response (ASMR) is getting more popular by the day. While this is not sound healing, it does use sound to elicit a physical reaction and deserves a mention. Like frisson (when you get goosebumps while listening to music), ASMR sends a pleasant tingly sensation across your skull and down your spine. You may have already experienced ASMR. It is similar to the feeling of someone whispering very close to your ear or, if you are old enough, you may remember when everyone went crazy for the two-levelled, long-pronged head massager.

While frisson feels exciting, ASMR triggers feelings of relaxation. Popular ASMR sounds, such as whispering, crunching leaves or tapping, calm your nervous system instead of stimulating it. The research into ASMR is very much in its infancy, but there is some discussion that it stimulates your vagus nerve to calm you down and make you feel better.

Some research at the University of Sheffield[61] found that people who watched ASMR videos and had a positive response to them had lower heart rates afterwards, suggesting that they were indeed having a parasympathetic response to whatever they were listening to. This study had some other interesting findings as well. Participants felt more relaxed, with lower levels of stress and feelings of sadness, and felt happier, less anxious and more socially connected.

It is important to note that not everyone is susceptible to ASMR and scientists do not quite know why some people have a reaction and others don't. If you are suffering from misophonia – an intense dislike and disgust of certain sounds, such as someone chewing – ASMR videos are best avoided, as these will be your idea of hell. If you do get the ASMR tingles, there are plenty of free videos out there for you to enjoy.

BINAURAL BEATS

Binaural beats are an auditory illusion. These are created in your brain when your ears hear two tones that have a slightly different frequency. One frequency enters one ear, the other frequency enters the other ear, and your brain takes the difference in the frequencies and creates a new, third sound from it – a binaural beat.

Because the tone is created from the small difference between the two tones you initially hear, the binaural beat has a low frequency. Where it is observed as a pulsing sound, the binaural beat encourages your brain to sync up with this low frequency – therefore taking you to a more relaxed alpha- or dreamy theta-dominant brainwave state. With that comes lower stress levels, reduced anxiety, a positive mood and increased creativity, as we discussed in Chapter 3.

Just as with drums, the benefits of binaural beats come from their ability to entrain you, which is why this is sometimes also known as brainwave entrainment technology. A sound practitioner can create these beats in a sound healing environment depending on their kit, but you can also enjoy them at home. Type in "binaural beats" into YouTube, Spotify and the like, and you will be greeted with hundreds of free audio. Just pop on some headphones and listen. Headphones are essential for binaural beats to work at home, as you need to hear each tone separately, one through each ear.

PROTECT YOUR HEARING

Have you seen any of those videos on social media where a sound is played and participants have to put down their hands when they no longer hear it? If so, have you noticed that it is always the youngster who keeps their hands up the longest? While some hearing loss will be due to natural and genetic factors, it is important to remember that when it comes to most hearing loss, environmental factors play a huge part.

Music concerts, bars and clubs have very loud music. I am not going to suggest you do not go to these, as life is for living, but when I hear someone's music from three seats away on a train, I know their earphones are clearly too loud. Just be a bit more conscious of the volume of music you are consuming, and be careful of noise, too. As a Londoner, I am always on the Tube, along with my decibel-reducing earplugs. Even one short 15-minute ride at over 100dbs is enough to potentially cause damage, so it is something to think about if you are doing it every day.

The same goes for sound baths: louder does not mean better. When I play sound, I am very conscious about staying within the parameters of what is healthy for the ears. You have to remember that not only are you hearing the sounds, you are also feeling them, and sometimes those deeper resonant sounds do not have to be booming loud for them to make a positive difference to your health. Playing too loudly also does not amp up the healing properties of a sound bath. Quite the opposite, in fact, as it could increase your stress levels.

I remember going to a gong bath many years ago where we were lying under a beautiful art installation. I was so excited and looking forward to the gong bath but within minutes I was feeling anxious. The sound was painfully loud and aggressive to my ear drums; my brain was saying "stop" and my ears were screaming it. I opened my eyes and saw that other people had also done the same and were shifting around uncomfortably. I turned to my side and put the blanket over my ear to try and muffle the incoming onslaught of sound waves.

A girl lying down next to me silently asked: "Is it supposed to be like this?"

"No," I mouthed back.

If you are a sound practitioner, sonic artist or therapist, please look around you. I watch my classes and my clients the entire way through, peaking up and around the gong if I have to, to make sure everyone is comfortable. You need to do the same and adjust your volume if you need to. Sometimes you may get carried away and the volume may end up a bit higher than you'd like, but this is not what I am talking about here – I am talking about having it sustained at an uncomfortable level. It will deregulate your clients if you play too loudly.

If you are a sound bath attendee, remember that not all sound has to be audible for you to gain the benefits. If you are at a live session where the instruments are there in front of you, there should be a natural ebb and flow with the volume. You are not supposed to hear it as if you were at a pop concert.

Sound Bite

Turn Down the Volume

Did you know that most hearing loss could be prevented and that getting old and losing your hearing is not something that goes hand-in-hand? The truth is, when most of us listen to music or videos while using headphones or earphones, the volume is too loud. Not only does this put your hearing at risk, but, rather shockingly, it also increases your chances of other medical conditions such as dementia.[62] I personally prefer using headphones over earbuds, which directly funnel sound straight into your ear canal. Whatever you choose, just keep the volume at an ear-friendly level.

CHAPTER SIX

"The secret of change is to focus all of your energy not on fighting the old, but on building the new."

– Socrates

How to Use Sound at Home

When I was first deciding how to write this book, there was one thing I was sure about – I had to include an audio recording. After all, how could I wax lyrical about all the benefits of sound and not give you some therapeutic sound to enjoy at home?

If you flip to page 239, you will find a QR code that will take you to this recording. This means that if you are a sound bath beginner, you can start your journey with therapeutic sound straight away. If you are a sound bath connoisseur, consider this a treat for you to continue your experience with sound in the comfort of your own home.

Because sound is a wonderful container to help with reducing stress, you can scan the code, close your eyes and sit on an armchair or lie on your bed and listen. The benefits will still reach you. This is the beauty and ease of sound healing.

But, with such a long career in wellness, having interviewed and spoken to the best health experts in the world and having written thousands of features on the best modalities out there for relaxation, I must share some excellent gems that will elevate your sound experience. Some of these practices I usually share only with my one-to-one clients, others make an appearance in one of my many group classes, and some I assign as exercises to private clients to carry out between their sessions. Finally, some I personally can't live without when I am listening to my instruments in my own home, so I am sharing them with you here, too.

Try out my recommendations, keep what you love and do not think you have to incorporate all of them. On some occasions, one will jump out at you, and on other days another practice will better suit your circumstances.

MAKE YOURSELF COMFORTABLE

Sound healing is about ease, comfort and rest. As we explored earlier, the first thing to do is set an intention. Sit for a moment and think about what you want from this sound bath. How do you want to feel at the end of it? What is calling you? What feeling or energy do you want to call in? Next, it is time to set your physical space.

Get comfortable, whatever that means to you. If you want to be seated and remain in a more engaged posture, this is totally fine. Personally, I am all about maximizing comfort, so if you can and want to lie down, then please do this.

When you are at home, your bed is always an option but lying on the floor on some cushions or a yoga mat is just as beneficial. I always recommend using a bolster or pillow under your knees to help support your back. You can even place a cushion or two under your feet to prop them up, too – this is especially nice if you have been on your feet all day.

Finally, keep a blanket handy or throw one over yourself before you start your sound bath. When you slip into a deep meditation, your body temperature drops and you are likely to feel cold.

POSTURE AND POSES

Most of us spend our days sitting at our desks or looking down at our phones, leading to terrible posture. Poor posture does not only *look* bad but it is detrimental to your health. When you are slouching with rounded shoulders and your head forward, you increase the amount of physical pressure on your spine. This weakens several muscles in your neck, negatively affects your breathing and compresses your vagus nerve – lowering its ability to send signals between your brain and your body.

Headaches, neckache, back pain and muscle tension are just some of the daily symptoms you may be facing. This is why, if you have made time for yourself to listen to some meditative sounds, it can be a good idea to use that time to also put yourself into positions that open up your body and give it a break from slouching.

If you are lying on the floor on a yoga mat, you could try some yogic asana poses. I like to incorporate these into my sound meditation practice. I am not a yogi, not even close, so it is important to honour and respect that these poses are borrowed from that practice.

Mritasana

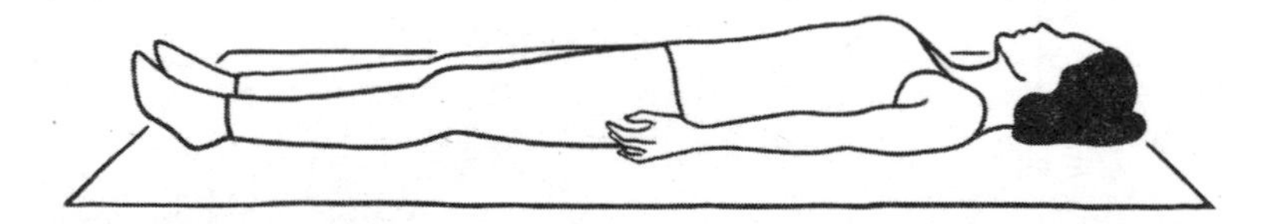

Mritasana is also known as the corpse pose or savasana. It is what you will have practised at the end of a yoga session and it may already be your go-to pose for any lying-down meditation.

- To do this pose, lie on your back and allow your arms (with your palms facing upwards) to drop open. Your arms should be at about a 45-degree angle away from your body.
- Your feet should be hip-width apart, with your toes facing to the sides. This is a wonderful way of keeping your body open to receiving.

Sound Bite

Smile

Keeping your legs uncrossed and your palms facing up towards the sky are examples of how to keep your body language open. Body language is important because how you physically express and hold your body reveals what is going on in your subconscious mind. But since this is a two-way street, you can use this to your advantage in a sound bath.

By using positive body language, you can trick your brain into thinking you are happier or more content than you actually are. Consciously tweaking your body language to add some positive adjustments can be so powerful that it can reduce your cortisol levels[63] and increase your happy hormones to boost your mood and fight off anxiety.

The mood-boosting effects of positive body language are all down to the facial-feedback hypothesis. Smiling engages certain muscles in your face, not only telling the *world* that you are happy but also telling *you* that you are happy – even if you are not. This is a real case of "fake it 'till you make it".[64]

To incorporate the facial-feedback hypothesis into your sound meditation, when you begin to ease yourself into a place of rest and comfort, follow these steps:

1. Allow your face to relax.
2. Soften your jaw.
3. Unfurl the space between your eyebrows.
4. Slowly lift the corners of your mouth.

Remember, you do not need to be grinning like the Joker for this to work: subtle is fine for a gentle mood lift.

Supta Baddha Konasana

The reclining butterfly, which is also sometimes called the "reclined goddess" pose in English, is deeply restorative. It is a great hip opener – perfect for those who sit at their desks all day as it helps to stretch out tight hip flexors. The hip flexors are the tiny muscles at the top of your thighs that shorten and tighten from being in a seated position for long periods of time.

- Lie down and follow the first step of the corpse pose (see page 171) with your arms resting beside you and your palms facing upwards.
- Bring your feet together and draw them closer to your body so that you can bend your legs at the knees.
- Now drop each knee to the side and bring the soles of your feet together, creating a diamond shape with your legs.
- You have the option to draw your feet closer or away from your body to make the posture more comfortable for you.

Gravity pulling on your knees will slowly help to stretch your groin, adductors (your inner thigh muscles) and hamstrings. But if your muscles feel particularly tight, you can place blocks or cushions under your knees to help support your legs. It is important to stay as comfortable as possible so you are not distracted from resting.

You can also place a cushion or ideally a neatly rolled towel under your neck to help support your neck better, too. If you wish to open up your chest even more, place a bolster or a block between your shoulder blades.

Viparita Karani

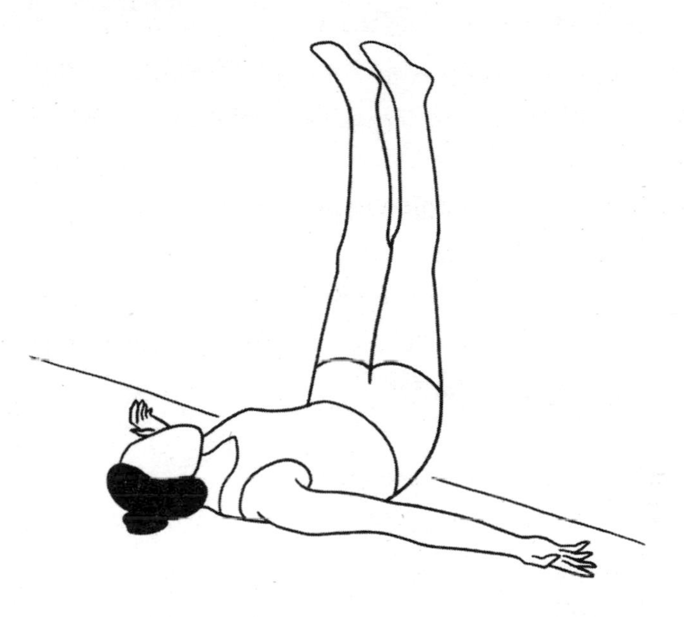

Viparita karani is another firm favourite of mine. You may know it as the "legs up the wall" pose. This gentle inversion asana is a great way to promote lymph flow, improve blood circulation and rest tired feet and legs.

- Sit as close as you can against a wall with your knees bent toward you.
- Lean back and shimmy your bum closer to the wall.
- Swing your legs up against the wall, stretching them out.
- Lie flat on your back and adjust yourself so you are comfortable.
- You can keep your arms away from your body or have them resting on you.
- Stay in this position for up to 20 minutes.

- When you are ready to release the pose, use your feet to push yourself away from the wall.
- Bring your knees to the centre of your chest and roll to either side, before moving into an upright position.

When you are in this pose, you can place a cushion under your head if this feels more comfortable. I personally love doing this pose if I am listening to a sound meditation just before bedtime. I also like to do this pose with an acupressure mat underneath my back.

Sound Bite

Embrace the (Acu) Pressure

Sometimes known as "bed of nails" mats, acupressure mats are foam mats with small plastic spikes. Working with the same principles of acupuncture (which involves inserting thin needles at various acu-points on your body), acupressure stimulates pressure points on your body without piercing your skin.

The spikes, perceived as painful or uncomfortable, trick your body into releasing mood-lifting endorphins. This helps you feel more relaxed, releasing tension from your body, and increases blood flow.

If you are lying on the acupressure mats as a beginner, it is recommended that you keep a thin layer of clothing between you and the mat so that it is not too overwhelming. You should also slowly build up the time you spend lying on the mat – beginning with a few minutes a day at first.

DEEP PRESSURE THERAPY

When you are in deep meditation, your body temperature drops and you may start to feel cold. This is why I always recommend doing any sound meditation work under a warm blanket. But there is another option – a weighted blanket.

These therapeutic blankets are inlaid with small weights sewn into pockets. They use the principles of a technique called deep pressure stimulation. The idea is that the pressure from the blanket mimics the feeling of being hugged and in doing so triggers your body to produce the love hormone, oxytocin. When this hormone floods into your body, it encourages your nervous system to relax.

Weighted blankets can be particularly helpful for people suffering from significant anxiety. More recently, studies have shown that the blankets can be useful for those in the neurodiverse community as well. When choosing a blanket, keep in mind that it should weigh only 10 per cent of your body weight.

People with claustrophobia may struggle to use one. If this applies to you or you simply do not like the idea of a heavy blanket, you can try another deep-pressure technique instead. While you are lying down, place one hand on your heart centre or sternum and one hand on your stomach. At the start of your sound practice, when you are consciously breathing more deeply, press your hands down gently with each exhale. Alternatively, you can give yourself a self-soothing tight squeeze.

NOTICE YOUR BREATH

The core of sound healing involves (re)gaining the ability to regulate your emotions and return to a parasympathetic, "rest and digest" state of being as your default. And you simply can't do that without also considering and potentially correcting how you breathe.

Every emotion is connected to your breath. It is fundamental to examine how you are breathing when you want to turn feelings of stress and anxiety into calmness. Think back to any time you have felt awful, with your stress heightened and an anxious feeling beginning to take over. I can guarantee your breathing was fast and shallow. Deepening and slowing your breath down signals the opposite.

THE IMPORTANCE OF NASAL BREATHING

One of my firmest beliefs when it comes to breathwork is that your mouth should be kept firmly shut. For years, there has been false belief that breathing is breathing and that it does not matter if it is being done through your nose or your mouth. I could not disagree with this more.

There is currently an epidemic of mouth breathing, especially in the West, and this is ruining our faces (long face syndrome) and our ability to regulate our nervous systems properly.[65,66] It is also causing a whole host of health issues, from incorrectly formed bites, increased allergies and postural problems to sleep conditions such as sleep apnoea. If you have a medical problem that is stopping you from breathing through your nose, please see your doctor. But for most people, mouth breathing is just a bad habit.

This is why at the start of every session, I tell my clients to breathe in – and, more importantly, to *breathe out* – through their noses. However, you can't change habits of a lifetime with one instruction in one session, which is why in group sessions I can immediately spot the mouth-breathers – they snore.

So why do so many of us use our mouths to breathe? The answer for the most part lies in our food. Most Western diets are soft, heavily processed and even have the fibre taken out (which is a whole other problem in itself). We are not chewing our food as we are supposed to or giving our mouths the chance to form correctly. This can leave us with a high, narrow palate that hinders our ability to breathe properly. Eventually, breathing using our mouths starts to feel easier. The easier it feels, the more we do it and it is not long before we are trapped in this unhealthy practice.

If you are a natural mouth-breather, can you do something about it? Yes, you can. By being conscious of how you breathe each day, you can make positive changes to improve your health, lower your stress levels and ensure you are breathing correctly all day and not just in your sound bath. Awareness of the problem is key. You can't change something for the better if you do not know you are doing it incorrectly in the first place.

One movement that is gathering a fair bit of traction is mewing, which was first created by an orthodontist named John Mew in the 1970s. At its essence, mewing is about correcting your tongue posture. By doing this, you then improve your breathing. Tongue posture is also important because when positioned correctly, your tongue activates your parasympathetic nervous system.[67] Meaning mewing is another way to encourage a full-body approach to relaxation in your sound meditation.

There is even a firm belief among the mewing community that it can also change your facial structure, giving you a more defined jawline, though the evidence is mostly self-reported at this stage. This means I can't guarantee you will develop an Angelina Jolie side profile from mewing, but knowing and maintaining correct tongue posture is important in helping you remain calmer in life, not just in a sound bath.

Stopping Your Mouth Breathing

To start to retrain your resting tongue position, you need to relax your tongue before suctioning it up against the roof of your mouth. This does not mean that your tongue touches the back of your front teeth – it should be pulled back a tiny bit and lie flat against your palate. If you have a high palate due to years of mouth breathing, you will not be able to achieve this fully, but rest your tongue as much as you can against the roof of your mouth and make sure the back of your tongue touches the soft palate. This may feel weird or even uncomfortable at first, but with enough practice you will soon get used to it. Sealing your tongue to the roof of your mouth also does something else – it signals to your brain via the essential vagus nerve (as discussed in Chapter 3) that you are in a resting mode, priming you for the start of a sound bath.

Another way to stop mouth breathing is by using mouth tape. Mouth taping is when you use tape to keep your lips together. Before you get carried away, this is not about taping your mouth shut completely as if you are in a hostage situation. You use a tiny bit of tape (you can buy mouth-taping strips) and attach it vertically to the middle of both your lips. The adhesive used is not very strong; if you need to open your mouth (especially if you are sleeping and are not aware of how you are breathing), the tape pulls away very easily and without any pain or discomfort.

The main purpose of mouth taping goes back to bringing mental awareness to a bodily function that you usually do without any conscious thought. Over time, you will no longer need the physical reminder of taping your mouth as it will become easier to nose-breathe. Eventually, this will become your default way of breathing.

If you find it too challenging to incorporate nasal breathing in your sound bath, then practise correcting your breathing techniques while you are in an alert and conscious state. I like to bring that awareness with me when I am exercising. I love spinning, so when I am on a bike I make sure that I only breathe through my nose the entire time. If I catch myself breathing through my mouth, I correct it immediately.

The key is to find that sweet spot between getting a workout and maintaining breathing with your nose exclusively. If you are gasping for air through your mouth because you have pushed yourself too hard, dial it back. Slow down your run, spin etc. to maintain the correct breathing "form", especially since the belief that you get more oxygen through your mouth when working out is not true. Long-term nasal breathing when you are exercising will improve your fitness as well as enable you to go deeper in your meditative sound sessions.

Feeling Nosy

I have been asked in sessions why am I so adamant about nasal breathing, so let me fill you in. Noses have evolved to be the perfect instrument to help you breathe as efficiently as possible and make the best use of the oxygen from the atmosphere. But this is not as simple as taking air from around you to go straight into your lungs. Being the first organ in your respiratory system, your nose does so much more. It is responsible for multiple functions, affecting the way you taste and even hear. It also filters and purifies the air to trap dust and harmful particles, which in turn will lower your risk of allergies including hay fever.

Your nasal conchae (also known as turbinates) are spiral structures that increase the surface area of your nasal cavity. When you breathe in through your nose, the thin hairs in your nasal conchae warm and moisten inhaled air, optimizing it for your lungs. While you are exhaling carbon dioxide, the trapped heat and moisture are being saved for your next inhale.

Nasal breathing also prompts your body to produce a compound called nitric oxide, which has antifungal, antiviral and antibacterial properties. Known as the miracle molecule, nitric oxide aids your immune system by protecting your body against many airborne pathogens and stops you from getting ill. Nitric oxide helps to open up your nasal passages, bronchi and bronchioles in your lungs to improve your lung volume so you get sufficient oxygen from each breath you take. It also widens your blood vessels to improve your circulation so that highly oxygenated blood is able to travel around your body more easily.

Improved blood flow to your arteries, veins and nerves combined with more oxygen in your system will improve your mood, increase your drive and help you feel calmer. Because your body is able to optimize each breath, your breathing rate slows down. This signals to your brain that you are safe, enabling your nervous system to calm down. Wider blood vessels will lower your blood pressure and heart rate, which can help to reduce feelings of stress, anxiety and panic.

Nitric oxide also stops you from being at risk of overbreathing and ruining the delicate balance between your oxygen and carbon dioxide levels. A common misconception is that the more you inhale, the more oxygen there must be whizzing around your body. This is not true – in fact, it is the opposite.

The more you breathe, the *less* oxygen gets delivered to your cells and tissues, because taking in too many breaths increases how much carbon dioxide you breathe out. Breathing out too much carbon dioxide means the oxygen in your blood stays in your blood and does not move into your muscles, organs and the cells where it is needed. (The medical term for this is respiratory alkalosis.)

What if it feels difficult to breathe using your nose? Unless you are ill (with a stuffy cold, for example), in which case it is just a temporary problem, for most people the more you use your nose to breathe, the easier it will become. This is a classic case of "use it or lose it" – the more you breathe in and out of your nose, the more you will avoid nasal congestion. You can also use sound to help fix your breathing. Regularly humming as described in Chapter 5 can help to release a blocked nose and improve your nasal breathing abilities. Try humming for five minutes each morning when you wake up.

BREATHE IN, BREATHE OUT

While breathing correctly will improve your health, certain breathing exercises will amp up the benefits even more. Here, I have included some of the best breathing techniques to use with your sound practice. I suggest doing one of these before embarking on each and every sound bath to help you feel calmer and less anxious.

You may find that once you have tried them all, you are drawn to certain breathing methods more than others or that a different one will resonate at a different moment. There are no hard and fast rules about only using one method each time. In fact, I often change what I use before each class or private session. If you have done some of these techniques before and been told that the exhale should be through your mouth, you will notice that here all the exhales are taken through your nose.

The Earth/Sky Breath

This is my favourite technique, which is maybe why I am putting it first. I discovered it in 2018. I was outside, barefoot on the grass, and the sun was shining, all of which undoubtedly helped to increase the mood-boosting benefits. But, years on, I still love practising this.

The Earth/Sky Breath is a full-bodied movement that can help you connect to your body and therefore pull you firmly into the present moment.

1. To begin, stand up with your feet shoulder-width apart.
2. Decide if you want to uplift your energy or feel more grounded, as this will impact on the order you do the next few steps.
3. If you are looking to feel uplifted, lift your arms up above you to form a "Y" shape. Take a deep breath in, looking up as you do so.
4. Slowly bring your hands down towards the ground, crossing them as you lower them.
5. Start to bend your body at your hips towards the ground, opening your arms out as if you are releasing your energy to the ground and breathing out while you do so.
6. With your lungs emptied, you should be standing in a half forward bend.
7. When you are ready to take your next breath in, raise your torso slightly while keeping your spine flat. Then inhale as you draw your arms together as if you are scooping up the grounding energy from the earth below.
8. As you return to fully standing position, you will find that your arms cross your body again. This time as you breathe out, open your arms to release that energy back to the sky as you are looking up.
9. Repeat the steps again.
10. If you are waiting to ground yourself, start your inhale as you are scooping up energy from the ground and end by breathing out towards the sky.

The Box Breath

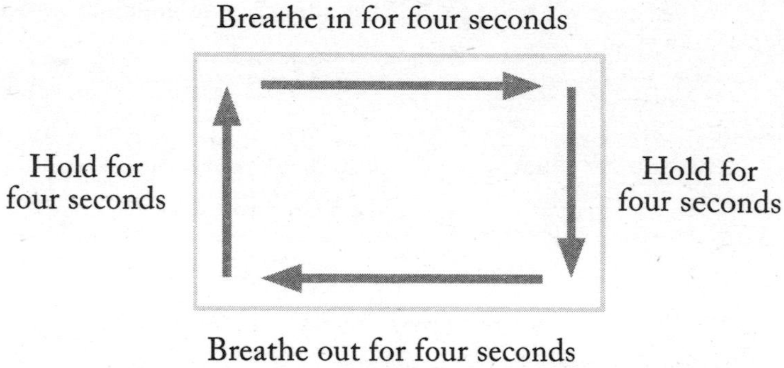

Breathe in for four seconds

Hold for four seconds

Hold for four seconds

Breathe out for four seconds

Many of you will be familiar with this style of breathing. It is such an easy one to do, especially before a sound bath or at any time you feel anxious or stressed. If you have trouble sleeping, you can try this before bedtime as well.

1. Start by taking a deep breath in through your nose for a mental count of four seconds, feeling the air pass through your nostrils.
2. With your lungs full, hold your breath for four seconds.
3. Slowly exhale, emptying out your lungs through your nose for four seconds.
4. Now hold for four seconds.
5. Repeat steps 1–4 for as long as you need.

The Double Inhale Breath

This is another firm favourite of mine. As the name suggests, it involves taking two breaths in before you exhale slowly. The double inhale breath is especially helpful if you have a habit of chronically taking shallow breaths, as it will help you discover just how deeply you should and could be breathing.

This breath is also known as the physiological sigh. Sighing is an essential reflex for healthy lung function and is something we all do naturally from time to time. You may sigh during your sleep if there is an excessive build-up of carbon dioxide (your waste gas) or if you are anxious or even upset and crying.

When you sigh,[68] you reset your breathing. Your lungs are full of millions of alveoli, which are tiny, balloon-shaped air sacs in which oxygen is swapped for carbon dioxide. Each time you inhale, your alveoli inflate. Sometimes, some of these tiny sacs collapse or dry out (from mouth breathing). The second inhale reinflates those tiny sacs, increasing your lung capacity and helping your body to remove carbon dioxide more efficiently from your system.

Sighing also helps to stimulate your phrenic nerve, which controls your diaphragm. When you reset your breathing, it also helps to activate the calming part of your nervous system. This is why it feels so good to sigh.

In sound healing, we view sighing as a form of release, too – helping to rid your body of tension and trapped emotion. And so, you may hear yourself or your fellow class attendees sigh during a session.

Sighing is a reflex that sometimes happens without your control. But when you intentionally sigh – or use this double inhale and long exhale breath that mimics the same bodily response – you trick your body into producing the same benefit. This breath will allow you to self-soothe into a less stressful state. It is the reason why I sometimes make clients sigh deeply or do the double inhale breath for two to three counts before doing any type of sound work.

1. To start, gently put your lips together and take a breath in through your nose. (Note: this breath will be longer than the second inhale.)
2. While holding that first breath, take another breath in through your nose – letting the air fill your lungs completely. This secondary breath should have also expanded your belly to the fullest.
3. Now slowly breathe out through your nose.
4. Repeat two or three times.

The Long Exhale Breath

This is also known as the 2:1 breath – you simply exhale for double the amount of time you inhaled for. Each time you breathe in, you engage your sympathetic nervous system, which keeps you alert and ready for action. When you breathe out, you stimulate the parasympathetic side of your nervous system.

When your exhale is longer than your inhale, your body understands that it is okay to relax and you are not in danger, tipping your body in favour of the rest and digest part of your nervous system.

1. To start, make sure your mouth is closed and your tongue is positioned against the roof of your mouth.
2. Take a deep breath in through your nose.

3. When your lungs are full, and without a pause, start to breathe out slowly through your nose.
4. To boost the relaxation response even more*, you can also engage your abs to gently help push out all the air as you reach the end of your exhale. Your belly will naturally expand on your inhale anyway, but the additional step of gently pushing the air out with your extended ab contraction means you are helping to increase your circulation, too.
5. Repeat as needed.

*When you breathe out, your diaphragm automatically moves up. Because there is less space around your heart, the pressure around your heart increases and this signals to your brain to slow down your heart rate, too – relaxing your body even more.

The 4-7-8 Breath

Once again, this breathing technique will help you feel more relaxed while also teaching you how to take deeper breaths. It shares many of the same principles of the extended breath but with an additional step. Holding your breath, as you also do during the box breath, will give your cells more time to absorb oxygen and produce carbon dioxide.

Simplistically speaking, carbon dioxide is the waste gas you breathe out but it also plays an important role in your body, so having low levels can cause problems. You need some carbon dioxide in your body to help release oxygen from your haemoglobin (the protein molecule that carries oxygen and gives your blood its red hue). These short bursts of holding your breath help to ensure you have the correct levels of carbon dioxide.

Holding your breath like this also helps to strengthen your diaphragm and improve your lung capacity, making it an excellent marker for longevity.

1. To start, assume the correct mouth position with your lips together and your tongue suctioned to the roof of your mouth.
2. Take a deep breath in through your nose for a count of four.
3. When your lungs are full, hold your breath for a count of seven*.
4. Slowly start to exhale (through your nose) for a count of eight.
5. Repeat for as long as required, until you are breathing normally again.

*If this feels tricky, start holding your breath for a count of five and release for a count of six until you can work up to the above.

Nadi Shodhana Pranayama

Also known as alternate nostril breathing, this is a technique taken from yoga. It has all the stress-relieving benefits that we have spoken about above.[69] The reason I have included it here is because I think it takes a little bit more concentration initially and can help bring you back into the present. Of course, the more you practise this technique, the more instinctual it will become.

1. While the rest of the breathing techniques mentioned can be done either sitting or lying down, this is best practised sitting. So get into a comfortable seated position, with your legs crossed, your back straight and your arms relaxed.
2. Bring your right (or dominant) hand up and assume the Nasagra Mudra hand position. This is when your index and middle finger are placed between your brows (below your third eye).
3. This leaves your thumb and ring finger (which is folded) free to close your nostrils as needed. (Your pinkie finger can be relaxed however you wish.)
4. To begin, exhale as normal and then use your thumb to gently close your right nostril (if using your right hand) and take a deep breath in through your left nostril.
5. Close your left nostril with your ring finger and release your thumb to open your right nostril.
6. Breathe out through your right side.
7. Inhale through your right nostril before using your thumb to close this nostril.
8. Open your left nostril and exhale through this side. This is one cycle.
9. Repeat for a few minutes.

Breathe Into Your Belly

A phrase I hear all the time when it comes to any breathwork before or after a sound bath is "breathe into your heart centre". I don't use this phrase as I feel that most of us are at risk of taking shallow breaths anyway. Imagining your breath only reaching your heart centre feels very high up in the body for me, and upper chest breathing is firmly where anxious people breathe.

Your diaphragm is just below your heart, but I feel that imagining yourself breathing into your belly enables you to take deeper breaths and really find out how much your lungs can take in. Correct breathing posture should see your belly move in and out rather than your shoulders moving up and down. This is why you will notice that with all these exercises I say you need to take your breath into your belly. Your diaphragm should be engaged with every breath you take. That act in itself will enable your body to function better and help to ensure your core muscles are working correctly, too.

To make sure you are expanding your belly, you are welcome to place your hands on your trunk when doing any pre-sound bath breathwork. This will help to give you a physical reminder to draw your inhale in fully.

Another really good way to bring an awareness to your breath posture is to breathe in a Makarasana pose, which is also known as the crocodile pose. Using this pose naturally encourages your body to breathe using your diaphragm and provides instant feedback (when your belly presses against the floor) of what you are doing.

Makarasana is a deeply restorative pose that enables your body to stretch out – great if you have been sitting at a desk all day. If you do this regularly, it will help to strengthen the muscles in your back, too.

1. Lie facing downwards so your belly and legs are touching the floor and your feet are relaxed.
2. Place your elbows on the ground* and rest your head into the palms of your hands so that your upper chest is elevated.
3. Relax your shoulders to lengthen your neck.
4. Breathe in and out so you can feel your belly contract and release against the floor.
5. Repeat for 5–10 minutes.

*You can also choose to keep your arms on the floor, resting your right palm over your left palm on the ground before putting your forehead on the back of your hand and closing your eyes.

Sound Bite

The Humming Boost

As I have mentioned, nasal breathing provides you with nitric oxide (among other benefits) and this is why I have insisted you do all of these breathing exercises through your nose only. But there is actually a sneaky hack if you want to turbo-boost your nitric oxide levels even further.

For techniques such as the box breath or the extended exhale breath, simply add in humming when you breathe out. This means when you do the exercises I have already described, you continue to breathe out through your nose while humming at the same time.

By doing this, you will increase the amount of nitric oxide in your body by an impressive 15 times[70] more than during quiet exhalation. This may even be as high as 20 times for some people and it is all down to the vibes. Scientists believe that the vibrations caused by humming increase air circulation and the capacity for the paranasal sinuses to create more nitric oxide.

MATTER OVER MIND

The good thing about these techniques is that they enable you to use your body to regain control of your mind. Sometimes when you are so dysregulated by chronic stress, anxiety, depression or even trauma, you simply do not have the capability to think your way out of it – to calm down by intellectually telling yourself that you are safe and that it is going to be okay.

While you rationally may understand you are safe in a particular moment and not being chased by a sabre-toothed tiger, your body is still signalling to your brain – overriding what your thoughts are saying – that you are not safe and will need to jump into action and neutralize/control/eliminate the "threat". When this happens, no amount of talking therapy and being in the "correct" mental space is going to help you rid yourself of feelings of tension and stress. This is why we have to come back to your body: you have to tune back into the physical essence of your being to regain control of your mental state.

How you breathe can be drastically altered by your emotional state, whether you are scared, anxious or angry or laughing, happy and relaxed. The reverse can be true, too – by changing your breathing, you can influence your emotions. Breathing correctly and engaging in these styles of breathing techniques begins to signal to your internal computer that you are okay to rest and slow down. This primes you perfectly to receive the nurturing sounds that will follow in a sound bath.

FEELING SCENT-SATIONALLY CALM

Another good way to engage your nose for a better sound healing experience is scent. Smell is a powerful, highly emotive sense that can instantly boost your efforts to induce a feeling of calm and peace. This is why spas tend to smell a certain way. Particular notes of vanilla, lavender or ylang-ylang will pop up again and again in spaces designed to help you relax, no matter where you are in the world.

Scent has a way of transporting you to a different place. Have you ever walked past someone on the street and their fragrance reminded you of someone or somewhere else? This is why estate agents recommend the smell of baking bread wafting around your home when buyers come to look around, and why smelling a combination of apples, cinnamon and pines may trigger you to think of Christmas.

When it comes to scent, noses are pretty spectacular. Not only can they detect over a trillion different smells, but aromas can impact on behaviours, too. Certain scents will help destress you. This is to do with how smell interacts with your brain and how your olfactory system (the sensory network that is responsible for smell) has an impact on your nervous system.

There is an evolutionary reason behind this. If someone in a tribe was scared and frightened, they released pheromones. The other members of the tribe could subconsciously smell the fear, helping them to become aware that danger was nearby. Before the advent of deodorants, noses would have expertly picked up pheromones in other people to help select the perfect mate, too.

When you sniff scents that please you, your nose tells your brain to release the happy hormone serotonin and even the reward hormone, dopamine. This means you feel calmer when you are enjoying certain essential oils and is why I love combining relaxing fragrances with therapeutic sound. Engaging more than one sense helps to take a client to a place where they can let their worries go.

Lavender, vetiver and chamomile, bergamot, sandalwood and clary sage are just some of the fragrance notes you can look for when choosing something to meditate with. Once you have picked something you like, try to use that exclusively every time you listen to a sound bath or meditate using sound at home. This helps to create a memory, an association, with your chosen fragrance. Because your olfactory nerve connects to the part of your brain linked with memory and emotion, the more often you use one particular scent each time you enjoy a sound bath, the stronger that link will become. This makes the process almost ritualistic.

We will look at rituals in more detail later on (see Chapter 7), but associating a particular fragrance with relaxation means that getting a whiff of that scent when you want to relax will also have a positive impact on your body. The scent highway works both ways. A spritz of a fragrance that makes you think of being calm and happy will signal to your brain that it is time to unwind, even if you are stressed when you use it.

If you are trying to decide which fragrance to use, play around with different possibilities to discover what you are naturally drawn to. You may end up with a scent you like in the day and another you like in the evening. Lavender is an obvious front-runner, especially if you want to relax before bedtime, while citrus scents can be used during the day if you are looking to feel more energized.

When a scent already has an emotional significance, it is easier to hack the command centre of your brain (your hypothalamus) to release calming hormones and increase your feelings of rest. Vanilla is universal. Apparently, it reminds us of our mother's milk, which is why it is so comforting and gorgeous to smell. Lemongrass will always hold a special spot for me personally. It reminds me of relaxing in faraway destinations where sea breezes are accompanied by the swaying of leaves in lush rainforests and where dark wood-laden villas hold whispers of gentle Thai voices.

Lemongrass also reminds me of spas and massages and of times when I have felt happy and calm. Because I have such a strong preference for that fragrance, it makes sense for me to choose it when I am doing deep meditative sound work. This is an example of why it is important to experiment with scents until you find the one that does the same for you.

Burn Baby, Burn

Full disclosure: not all candles have been created equally. In terms of indoor air pollution, candles are not always the healthiest choice. Most are made from poor-quality ingredients, and so do not burn very cleanly. However, candles give off a wonderful warm, yellow glow. These days the battery-operated versions are surprisingly good and they even flicker. So if we are talking lights, those are a great option. But here we want to use candles for their scent.

If you are going down the candle route, choose soy ones, as these provide a cleaner burn and require a lower temperature – meaning they last longer, too. Shop around to see what the candle is made of and do not be fooled into thinking the more expensive it is, the healthier the candle will be. Some of the worst candles I have tried in the past are expensive ones from luxury department stores.

Once you have your candle, treat it correctly. This will not enhance your sound bath but will make sure you are getting your money's worth from your waxy purchases. The first burn is crucial and will determine how the candle behaves and how long it lasts.

- Trim the wick before every burn. If it is too long, you will get a flame that is too big and has a sooty burn.
- Since most scented candles are in jars, it is important to let the wax melt completely across the entire surface of the candle on that first burn. This gives your candle the correct wax memory – avoiding the dreaded tunnelling that can happen otherwise.

Sound Bite

Fixing Tunnelling

If you did not have time to burn the candle fully the first time and it has started to tunnel, you can fix this – it will just take some patience. Grab the tunnelled candle and wrap some aluminium foil around the edges. The foil needs to have an overhang into the candle but leave a hole big enough to not conceal the flame. Then light your candle.

The reflection of both the flame light and heat on the solid edges of the candle will slowly enable it to soften and melt. This will even out the top layer of wax and stop you from wasting most of the candle.

Flame-Free Alternatives

If you already have several scented candles, or the fragrance you want is only available in a candle that does not burn cleanly, you could try a scent warmer. Also known as a melting wax lamp, a scent warmer heats the candle wax from above (or the sides) so that it slowly releases its fragrance without being lit. Since there is no flame, there is also no smoke or soot. A lack of flame means that you can rest fully or even fall asleep straight after (or during) your sound bath without worrying about accidentally burning your house down. No flame also means your candle will last longer.

Candles are not your only option. These days the market is full of room mists, relaxation sprays, visually pleasing diffusers and incense sticks. Research shows that essential oils have antibacterial, antiviral and antifungal properties, making them a tool for so much more than just de-stressing. I have a few favourites and often use an air spray that is packed with sandalwood, lavender and neroli. There are also many sleep-science backed sprays. These can be excellent options for your sound baths, too.

Another room spray I love has hypnotic notes of citrus and clove and this brings me to another important point. Aim for something you adore, because if you are excited to spray it, you will use it. Make sure it is comfortably within your budget as well. You do not want to buy something that you feel guilty about using and therefore try to "save for best" and never use.

SMOKE CEREMONIES

If you want a more ceremonial start to your practice, a smoke cleanse may be the way to go. Frankincense is made from the sap of the Boswellia tree and has been used for centuries to help clear away anxious thoughts and lower feelings of stress. When used in resin form, it is easy to burn and will provide ample smoke while releasing a scent that feels extra special. If you are keen to try Frankincense, you will need:

- pieces of Frankincense resin
- charcoal discs
- mini tongs
- matches / lighter
- a heatproof dish

And here's what you'll need to do:

1. Set out your heatproof dish and make sure the lid is open if it has one.
2. Holding the charcoal disc with the mini tongs, light the charcoal.
3. Place the disc in the heatproof dish and wait for it to turn grey (which means it has stopped burning).
4. Place a piece of resin in the middle of the now-grey charcoal disc. If the space you are cleansing is small, pick the smallest piece you have. The bigger the piece, the longer it will burn and the more smoke it will produce. Another tip is to cool the charcoal disc for up to 10 minutes before you add the resin. This will give you a manageable smoke.

Avoiding Cultural Harm

We have a responsibility to respect indigenous cultures and their practices and Mother Earth where possible. Palo Santo wood smells divine but there is confusion about whether or not the *Bursera graveolens* tree (from which the aromatic wood comes from) is endangered. It is agreed, however, that the trees' habitat is rapidly shrinking. Western interest has also skyrocketed the demand for Palo Santo, which means the wood is being overharvested and harvested incorrectly.

Palo Santo should be collected from trees that not only have died naturally but have been left to rest on the forest floor for several years. This enables the wood to fully develop its aromatic (and therapeutic) qualities. Even if the sourcing is ethical, the wood supplies are highly limited and these trees are sacred to their communities. So unless Palo Santo is a direct part of your ancestral heritage, I would recommend you do not use it.

White sage is sold in bundles that are often wafted around a room to cleanse the energy of a space, but nearly every non-native person who embodies this practice does so incorrectly. When burning white sage, you should only burn one leaf at a time. The belief is that lighting a bundle would actually clear a room or space of *all* energy, according to community elders.

Communities indigenous to the Americas also see burning white sage as a "closed practice", meaning that you should not be dabbling in it if it is not within your heritage. As recently as 1978 (when the American Indian Religious Freedom Act was introduced in the USA), it was illegal for First Nations communities to practice this scared ceremonial act (along with a whole host of other religious and culturally significant rituals). In fact, these communities are still fighting for their right to use white sage ceremonially in hospital for their loved ones.

Fortunately, our own countries are abundant with alternative herbs and materials that can be used instead, providing all the benefits without the cultural harm.

- Swap white sage for clary sage – clary sage has a calming effect on your body and helps to ease feelings of stress.
- Use cedarwood sticks instead of Palo Santo.
- Try garden herbs, such as rosemary. Cut some sprigs and dry them in a low oven or out in the sun (if you are lucky enough to have good weather). Once these are fully dried, tie them together to create your own bundle.

My friend Maria, a UK-based acupuncturist, herb enthusiast and fellow sound practitioner, also adds the following wisdom: "It's nice to use something local. You can set an intention from the moment you begin picking the herbs. Herbs grown locally, grow locally for a reason. They support our constitution in the climate where you and the herb are from."

I love this idea of using local herbs to connect you to your environment.

LET THERE BE - DIMMED - LIGHT

Setting your space before a sound bath or any meditation is crucial and while you can forgo some things, this is a non-negotiable one for me. In fact, one particular rule I live by is that the main overhead lights are never to be used at home. There is nothing relaxing about a 100-watt bulb burning brightly above you at the best of times, so it certainly can't make an appearance when it is time to wind down.

Lighting and different hued lights have a huge impact on our entire physical body. By now most of us know that we should not look at a phone in the evenings before bedtime. The phone's blue light messes up our circadian rhythms and our sleep quality and quantity. But did you know that blue light is also found in LED lights and overhead halogen bulbs?

We all love the warm reds, oranges and rich pinks the skies can turn during a sunset. We may not understand why this is so relaxing, but we just inherently feel that it is. Bright artificial lights (especially white light) signal activity – that it is time to engage, move, act and do. Warmer lights on the shorter end of the light spectrum tell your body that it is time to relax and just be – to slow down and become still and to surrender to a more peaceful state. These lights signal to your brain that it is okay to down tools, rest and recover.

For this reason, lighting is so important when it comes to preparing your space for a sound bath. You can have the warm glow of candles, real or battery-operated, but all types of soft lighting will work. A well-positioned lamp can give your space an inviting glow, while fairy lights can add a touch of magic. If you want to add colour, a Himalayan salt lamp will provide a pink-red richness. A sunset lamp, which I sometimes use in sessions, will give you a diffused rainbow of soft colours. A galaxy projector with stars dancing on your walls and ceiling will add a playful, whimsical touch.

If you are doing a sound bath during the first half of the day and you want to feel more energized after your session, it might be a good idea to let the natural light from outside seep gently into the room you are in, as long as it is not too bright or uncomfortable for your eyes.

Using Eye Masks

If you want to completely shut out the outside world or prefer total darkness, eye masks are the way to go. Fortunately, you can choose from an array of comfortable options. Soothing heated masks will revive tired eyes. You can either buy disposal ones that are individual wrapped and self-heat, or a wheat-filled one that you pop in the microwave for a few seconds to warm up. You can also buy comfortable padded silk masks that have wraparound thick bands to block all light.

DITCH THE PHONE

Imagine the scene . . . You have set the perfect environment, played beautifully and seen the client fall into a deeply meditative state. When you bring them back from the serene meditation they have experienced, they look rested and have been able to successfully turn off. Just as you let them know they can leave, their face is illuminated by the blue glow of their smartphone. It feels as though all your hard work has been undone.

Blue light from your phone not only messes up your sleep when you are scrolling before bedtime, but in recent years there have been more discussions about not looking at your phone first thing in the morning either. Just before you wake up, your brain slowly starts to move into a theta brainwave state from its nighttime delta brainwave state. As discussed in Chapter 3, theta is when you are in a deep dreamy state. From there, your brain slowly moves into a relaxed alpha brain state. This is when you are awake but not concentrating, so your mind and therefore your body are still relaxed.

Imagine it like this . . . You are at the bottom of a staircase and have steep steps to climb before you can start your day. You slowly move up each step to reach the top. Your brain is the same: it moves through each phase to set you up for the day.

Now imagine you are being chased up the stairs with someone snapping at your heels. Your sense of urgency may either cause you to trip up the stairs or make you feel stressed and scramble up. Chased or not, the end result may initially appear the same: you have made it to the top of the staircase. But it is important to remember that those who are chased up the stairs will not be at the top feeling calm and steady. Instead, they will be anxious and unsettled.

The same goes for your brain. Scrambling or skipping through the brainwave stages that *should* happen leaves you unsettled, and that feeling of being chased (anxiety) may even stay with you for the rest of the day.

The same principle applies when you "come to" from a sound bath experience: you need to slowly move though the different brainwave states. In the same way you experience REM sleep, a sound bath can take you to a theta brainwave state. Grabbing your phone immediately and checking your emails or notifications forces your brain to skip these important stages – hurtling you immediately back to the most awake, alert (beta brainwave) state.

Instead of reaching for your device, give yourself a few minutes to continue resting. Stare into space, curl up on your side, sip a warming cup of herbal tea and just be. There is no correct amount of time you should ditch your phone. But if you can gift yourself (at least) 30 minutes without it, then do so. I also recommend turning notifications off your smart watches before a sound bath.

Sound Bite

Do You Need Your Own Instruments?

This is another question I am often asked, and the short answer is no. Someone will come to their first sound bath, feel deeply rested and in their excitement decide that they want to buy a sonic kit. I do not think this is necessary. When you are the one playing, you are in a thinking space with a concentrating mindset. When someone is playing for you, you just get to *be*. You are in a state of receiving. This means that fundamentally the experience is different. If it was not, there would be no need for sound practitioners to attend sessions as clients. But we do. A sound experience is not just about hearing an instrument being played, it is also about letting go and having someone hold space for you. Perhaps the only exception to this rule would be wind chimes. You can hang these up and let the sounds come to you whenever the wind moves them gently.

REAPING THE BENEFITS

As this chapter ends, I want to reaffirm that you do not need to carry out all of these suggestions for every sound bath. Combining and adding one or two of these extra modalities, as and when you need them, will enable you to get even more from your sound bath experience. It will amplify the results and cement your spot in a place of rest, elevating your experience. You can also choose to enjoy the sound with your eyes closed and nothing else. The benefits will still be there and you will receive what you need from it.

CHAPTER SEVEN

"Gratitude is not only the greatest of virtues, but the parent of all the others."

– Marcus Tullius Cicero

Mixing Modalities

Sound by itself is a beautiful container for growth, healing and transformation. As we have explored throughout these pages, its benefits are wide and varied. Sound is undoubtedly one of the best tools you have to create positive change and increase the potential for peace in your life. But it is not the only healing tool out there. Fortunately, my years spent exploring, testing, investigating and practising other modalities have meant that I have been able to curate a list of activities that I believe work in perfect harmony with sound healing.

All the modalities in this chapter are "sound adjacent" and have the ability to boost the benefits of your sound healing experience. I recommend some of these to my clients at the start of their healing journey, while others work better once clients have built up their sound practice. You may read these pages and jump on an idea you want to add to your life immediately or decide you will come back to these suggestions in time.

Some of these practices will incorporate your other senses, and others will strengthen the concepts of resonance (from Chapter 3). My aim is to get you to a place of rest and restoration so that you can deplete any chronic stress and anxiety from your day-to-day life. And these recommendations are there to help provide you with as much additional support as you need to get you there.

There is a wealth of activities and modalities and I do not have the space to cover all of these in this book. Everything I have chosen for the following pages works particularly well with sound healing. I have tried and tested them with my own clients to ensure they give the best results.

EXPRESS GRATITUDE

Being thankful for everything you have is a cornerstone of having good mental and emotional health. I stick to this pillar firmly in my own life. Alongside sound healing, gratitude has helped me during tough times and brightened up otherwise normal days. It takes no time and its benefits are sizeable. You can add gratitude into your thoughts on your way to work, during your supermarket shop or when you are gently rousing after a sound bath.

We live in a society where comparison dominates and it is stealing our peace and joy. Focusing on what you may "lack" and what others have will keep you in a place of resistance. Your brain will keep seeking to validate a scarcity mindset and that abundance is not for you. Practising gratitude will help to shift that mindset and open you up to more fulfilment. It literally rewires your brain on a biological level by changing your brain chemistry. Paying attention to what you have prompts the release of a cocktail of feel-good chemicals in your brain, such as serotonin, dopamine and norepinephrine – these are the neurotransmitters that help to manage your emotions and immediate stress responses. This makes you feel happier, more satisfied and less anxious.

Just as sound healing encourages you to see more resonance and more positivity in your life, spending time recognizing how good your life is leads you to becoming even more thankful. This positive reinforcement strengthens the neural pathways in your brain. Over time you feel content with what you have because you focus on the good, moving you away from any negativity bias you may have to a more peaceful and nourished state of mind.

Research published in the *Journal of Personality and Social Psychology* found that just 10 weeks of writing about what you are grateful for makes you more optimistic about life and encourages you to exercise more regularly.[71] You may even become ill less frequently than people who write down their daily frustrations. A further study from the Greater Good Science Centre at the University of California Berkeley found that within eight weeks, MRI scans show a stronger brain structure for social cognition and empathy in people who consistently give thanks.[72] Meaning that being grateful changes both your perspective on the world and it makes you a better person, too.

When you are in a sound healing session, you are tied to the present. You can't be doing anything else. Your eyes are closed, you are lying down, but you are not asleep. This is the moment to truly embrace that you are a human *being* not a human *doing*.

Gratitude also gives you a strong awareness of the present, which is another reason it works so well with sound healing. You can't worry about the future or mull over the past when you are rooted to the present moment and all the good that you have. When you finish a sound bath, your mind is perfectly prepped to access that healing space of gratitude. As I have mentioned throughout this book, sound leaves you feeling rested, and it is always easier to see happier moments, embrace joy and tap into positive feelings from a place of relaxation.

One stumbling block about showing gratitude, especially if you are new to the concept, is to reserve it only for the big moments. But you can't run out of gratitude – it is limitless. You do not have to save your thanks for life-changing events such as big purchases and milestone news. It is easy to be thankful when you have a new car, job, home, designer handbag or holiday booked. But seeing joy in the smaller interactions will change your life in a more expansive way. Giving thanks to those tiny moments, the ones that could easily go unacknowledged, is where the magic lies.

This does not mean you have to be thankful for every moment of every day. It is is about recognizing that each day, even the sad, unhappy or stressful ones, have good moments. You may be grateful for the multitude of spectacular colours a sunrise provides, a crisp sunny day in winter, the smell and taste of your favourite coffee, or unexpectedly bumping into a friend on your commute. Sharing a conversation with a stranger at the bus stop or witnessing a sweet interaction between a parent and their child are pockets of joy in every single day. And while you may physically see them, you need to acknowledge and mentally register moments like this to *really* see them and reap the benefits.

You will notice that I have included events or situations that may not happen to you directly. You may wonder why you should feel gratitude for witnessing a sweet moment between two strangers. In its essence, gratitude is about flipping a switch on how you see life. So just bearing witness to someone else's joy can also be a moment for you to be thankful. We do this with good friends already, being happy for their joy when they announce a new job or promotion, or on their wedding day. Since gratitude is not transactional, there is no reason why you can't extend a feeling of happiness on seeing someone else's contentment.

Many a time I find myself being grateful for the energy people bring into my sessions – both private or group. I give thanks when I see regulars coming in for a treatment or when someone decides to try sound healing for the first time.

Your gratitude practice does not have to be solely external and you can and should include things you did as well. Look inward to say thanks, too. Perhaps you assisted someone in their day, gave up your seat to someone who needed it more on your commute or let a visibly stressed parent skip the coffee queue. In these moments you can say thanks for having the opportunity to be kind and fortunate that you were able to make someone else's day better, easier and more comfortable. You can also thank yourself after a sound bath, expressing gratitude that you had the opportunity to let rest-inducing, healing sound waves wash over you. If you are at home, be thankful that you can have a sound bath in the comfort of your own room or house.

KEEP A GRATITUDE JOURNAL

You can use any notebook you already have or pick up something new. It does not need to be expensive or fancy. The important thing is to keep your thankful notes together so that you can revisit the pages when you need to.

- Before your sound bath, jot down three things that you are grateful for. These can be as simple as, "Today I am thankful I got to walk through the park on my way home."

- Once you have written down your *what*, add a line or two about how this makes you feel. This is your *why*. For example, "Today I am thankful I got to walk through the park on my way home. Doing so gave me the chance to enjoy how beautiful the trees looked."

If something surprising happens, highlight this in your gratitude journal, even if it did not happen on the same day as your sound bath. Because these interactions catch you by surprise, you will tend to have stronger levels of gratitude towards them.

Remember that each gratitude does not have to be dramatically different. Do not put pressure on yourself – no one else is going to read your journal. Over time, you will find the words start to flow more easily.

SEND A GRATITUDE LETTER

This activity is a good way to recognize and appreciate those around you and stop you from taking them for granted. Sending a letter sends out ripples of positive feelings. You feel good writing the words and the recipient feels happy reading them. There is something special about receiving a physical letter or card so do not opt for an email or text. This will also help you stay away from your device, where you could become distracted by notifications and to-do lists.

You can choose to do this exercise before or after your sound bath. Think about someone who has made a positive impact on your life. Someone who perhaps has inspired you or is just fun to be around. Write down why you are thankful for them and the ways in which they make your life better. Pull out one or two examples of how you felt and how much you appreciate them – expressing what they mean to you.

HAVE A TA-DAH LIST

Another way to implement gratitude into your day is by celebrating little wins. Ticking items off to-do lists lets you feel satisfied. But to-do lists (while very necessary at times) can also make you feel as though you have never caught up with everything that needs to be done, especially if there is still a lot left for you to do at the end of the day. The Ta-Dah List is the fun, positive alternative. It helps you see everything that you have accomplished.

- Compile this list throughout your day so you do not forget anything. (You can still have a to-do list for practicality; just put it on a different sheet of paper to your Ta-Dah list.) Each time you add to your list, your brain will reward you with a dose of the happy hormone dopamine.

- Before your sound bath, take the time to sit with your Ta-Dah list and see everything you have achieved – giving thanks for your positive actions. This is an easy way to shift your focus.

THANK YOUR FUTURE SELF

This is another great way to start changing your mindset. Pen a letter to your future self, thanking them for all the things you have or will accomplish, such as: "Thank you for sticking to my weekly sound baths; it helped me refocus my goals." Once again, write as much or as little as you like. You can write in your journal or on a separate piece of paper and seal your letter in an envelope to read in six months' time.

EXPRESS YOURSELF THROUGH ART

It goes without saying that I am an advocate of engaging the senses when it comes to releasing stress and combating anxious thoughts. But your sense of hearing is not the only way to do this – you can use your eyes, too.

While it is not mandatory to close your eyes during a sound bath, most people do. Doing so relieves you from your visual dominance and enables you to enter a state of quiet wakefulness where your body recognizes that you are safe enough to close your eyes. However, engaging in your sense of sight before and after a sound bath can be very therapeutic.

Art is an excellent tool for both awareness and mindfulness. Just like sound, it does not require the use of words. Whether you are drawing or painting, you can express yourself in a way that transcends verbal communication. This may help you sink further into a space of exploration. One of the best things I love about sound healing is that you can access your emotions and experiences in a session even if you can't articulate what is making you stressed, anxious or unable to rest. Therapeutic art works on a similar principle.

In Chapter 3, I explored how sound healing can help you to tap into your creativity, which is why engaging in a creative outlet after a sound bath works particularly well. Your brain is still in a relaxed, slower brainwave state, a time when you should stay away from your phone. If you can't trust your idle fingers to stop reaching for your device, pick up a paintbrush instead. This will prolong the time your brain is in a slower brainwave state after your session – increasing the benefits.

How you express yourself artistically is up to you; it is more important to figure out what works best for your circumstances. Here I have included some of my go-to exercises. Remember that you do not need to be a proficient artist to do this. Technical skills are not a requirement; this is simply a creative exploration of your emotions. Just make sure you have everything ready before your sound bath as there is nothing relaxing about raiding your house in a panic to find supplies.

FREE-FLOW DRAWING

Drawing and painting help you stay in a relaxed, alpha-dominant brainwave state. Choose whichever medium you like. Once you have finished your sound bath, allow yourself to stay without thought and let your hands embellish the paper or canvas in a way that suits you. If writing is not your thing, this could be a better alternative.

PAINT YOUR EMOTIONS

Before a sound bath, think about how you are feeling and then draw, sketch and scribble your emotions. Which colours are you drawn toward? Which paint strokes or shading style feels right? Allow yourself to feel your emotions before your sound work. Then, after your sound bath, come back to your artistic expression. You can use the same piece of paper or a clean sheet. Which colours are you drawn to now? Has it changed? Which emotions are you feeling now, and how would they look if they were painted or drawn?

Remember there is no right or wrong here. This is just another tool for your self-expression, to help you connect better to your state of being.

COLOUR IN

A few years back, you couldn't take a trip to a bookshop without encountering adult colouring books, and for good reason. Colouring is a mindfulness technique that helps you switch off and unwind. The repetitive action of colouring is calming, predictable and easy and is known to help soothe feelings of stress.

Colouring can be helpful to do before a sound bath if you are feeling particularly anxious as it will draw your attention to the present moment. Then, once you are ready, you can put your pencils down and listen to the soundscape I have recorded for you.

CURATE A VISION BOARD

I am a firm believer that if you can think it, you can do it. Seeing a visual representation of yourself doing something that may be challenging helps to reinforce the belief that it is possible. This is why sports personalities will often say how they imagined winning a championship or medal. Making a vision board is a great way to hone in and visualize what you want.

- Grab some magazines or brochures and cut out the words and images that you are drawn to. You are also welcome to print off images you find online.

- From your clippings, glue together how you want to feel. Do images of turquoise seas and pristine beaches signal rest and relaxation? Maybe the dense jungles of South-east Asian countries invoke a sense of calm for you? You can do this before or after your sound bath. If you are creating a vision board before a sound bath, you can spend the duration of your sound experience thinking about your creation and how it would feel to live your dream life. If you are putting something together after a sound bath, your rested mind may find it easier to decipher what would make you happy. Perhaps the answer may even come to you during your sound bath.

AFFIRMATIONS

Affirmations are positive words or phrases that help to rewire your inner dialogue. This concept is the linguistic embodiment of resonance and was created in the 1970s by two neuroscientists. Reciting positive phrases helps to shift your mindset and tells your brain that you can achieve or gain something. The more you repeat that positive phrase, the more your brain seeks to confirm what you are saying.

Your brain can't distinguish between what is real and what you think is real. The negative impact of this can be a heightened sense of anxiety. But affirmations can help you use this process to your advantage and counteract this response. If you are feeling stressed about a particular event, such as an interview, rather than thinking about it from a place of fear, use a positive affirmation such as: "I love how confident and self-assured I feel." This will help you acknowledge that you have all the required skills and experience to ace the interview.

Affirmations do not just work for specific situations; you can also say them to welcome general feelings of calm across your daily life. This is where using affirmations *after* a sound bath can step into play. Your internal belief system learns from your subconscious mind, not just your conscious one. While affirmations can be used in the middle of a highly stressful situation, you can't lay the foundations for them at this time. Just repeating "I am calm, I am calm" will not work because your brain will not believe you. Your body will have already dispensed stress hormones so it will be too late and your words in that moment will not match up to the signals from your body.

For affirmations to work when you are feeling stressed or anxious and get you back to a place of calm, you must first teach your brain how you *feel* when you are calm. You need to match the affirmation to an experience. Affirmations need to have some basis of truth so that your brain can connect the right neural pathways together.

For most of us, long-term stress and anxiety is firmly in our minds regardless of what the signals from our body say initially. Our mind then sends signals down to our body to respond as if we are in danger. What we are doing here is using the bottom-up approach – body first, mind second – to confirm a change of pace.

Repeating statements of being relaxed and at peace when you feel relaxed and at peace (such as after a sound bath) is the perfect way to reinforce that message. Allow your brain to wire the words to the feeling when the affirmation is true or close to the truth. This validates your words and the experience.

Showing your brain that the affirmation is correct enables you to then access it when you are unnecessarily overwhelmed or anxious. This is why saying affirmations after a sound bath works so well. Your subconscious mind is prepped by the sound waves to receive what you are telling it. And since you will still be in an alpha-dominant brainwave state, you are more open to suggestion and the message is more likely to stick, so take advantage of this.

Sound Bite

Is a Mantra the Same as an Affirmation?

A mantra is a sacred prayer, phrase or chant used in meditative practice in both Hindu and Buddhist traditions and has religious and spiritual importance. The pronunciation of the words is believed to hold vibrational qualities to invoke their power. I have often heard mantra and affirmation being used interchangeably, but these are different and serve different purposes.

Repeating a Sanskrit mantra that does not hold any significance, importance or understanding for you will not have the same impact as creating your own affirmation. However, mantras such as *om namah Shivaya* (which means: *my salutations to Shiva, the auspicious one*) can be incredibly powerful if these are within your understanding.

YOUR 5 STRESS-BUSTING AFFIRMATIONS

To boost the power of your affirmation, make sure your words are in the present tense and speak them as though you already have achieved the desired outcome. So "I *am* calm" is better than "I *will be* calm". Doing this engages your reticular activating system once again. It will help you see the world as you wish it to be.

Here are five affirmations I recommend you use:

- "I feel calm and I am happy."
- "I am safe and I am well."
- "I inhale peace and exhale worry."
- "I am capable of handling whatever comes my way."
- "I am doing the best I can and that is enough."

Alternatively, pen something that feels true for you, remembering to keep it short and memorable. Repeat one of your affirmations several times after your sound bath.

CHANGE YOUR PASSWORDS

Think about the number of times in a week or month you may type a password into your office computer, laptop or phone. It adds up, doesn't it? Most of us type in meaningless combinations of phrases and numbers but this simple trick is a useful way to add *general* affirmations into your day without any effort. Why not change your passwords to positive personal statements?

Something simple such as, "Good things happen to me", "I welcome in abundance", "I deserve good things" or "I am thankful 4 everything" can work well here. You will recite it in your head each time you type it in – reinforcing helpful self-talk throughout the day or week. Doing this does not add any extra time to your day and it is far more beneficial than "Password123".

WRITE IT OUT

I have previously mentioned how keeping a gratitude journal is beneficial, but therapeutic, free-flow writing is another helpful tool for self-discovery and transformation. Long before I embarked on a career as an editor (let alone a sound therapist), I always found that whenever I was feeling anything too intensely, I would write. When I couldn't organize my thoughts verbally, I would write. Pouring your thoughts out onto paper is deeply cathartic.

The beauty of sound healing is that it enables you to tap into your innate wisdom. Sound healing shows you how you are feeling even when you do not yet have the conscious vocabulary for it. But once you have accessed or started to access your emotions, journaling can be a useful way to further understand your previously concealed thoughts.

In a sound healing journey, you may experience an ASC. If you do, write it down while the memory is fresh. What did you see? Which thoughts, images and memories came up for you and what could they mean? Are the meanings obvious or do you need to explore them further? Writing it down could make everything clearer.

Another reason that journaling is a good activity to do after a sound bath is something we have discussed already. As you know by now, a sound bath changes your dominant brainwave state. If you have reached an alpha- or theta-dominant brain state, this does not change the second your sound bath ends. You remain in an alpha state at the end of your sonic journey where you are mentally aware but physically relaxed. I mentioned in Chapter 6 how you should not disrupt this time by looking at your phone. Another alternative to art therapy and expressing gratitude (modalities mentioned in this chapter) would be to spend your time penning your thoughts.

If you are completely new to writing in this way, it can feel daunting. And that is okay. Simply doodle at first, just getting your pen to paper (once again, this is a tech-free zone). You can even start with "I don't know what to write . . .".

Let your first words be a stream of consciousness. Whatever pops into your mind, put it into ink. If you get into the habit of writing, the rest will come. Also remember this is just for you. No one will read your words or see them so you do not need to be coherent or use neat handwriting. It is also worth noting that after a sound healing session you are in a flow state, so you may even find that your thoughts travel through your pen with greater ease than you had anticipated.

You can journal before a sound bath as well, but this will serve a slightly different purpose. Journaling can help you to recognize your fears and self-limiting beliefs. Writing down your inner thoughts forces you to slow down and helps sort through how you are feeling, organize your emotions and recognize what your personal triggers may be. This enables your amygdala (the emotional centre of your brain) to better understand your emotions and not overreact to experiences, such as recognizing the difference between a harsh email from your boss and a sabre-toothed tiger on your trail.

Essentially, free-flow writing improves your emotional intelligence and, if you are feeling stuck, can help lead you to a path of illumination. Taking this knowledge into your sound bath will help you focus on what you really want to shift in your session.

JOURNAL PROMPTS TO TRY

Ask yourself more inquisitive, probing questions before a sound bath, and then afterwards let your mind remain in a place of resonance and creativity. You only need to answer one question each time, so allow yourself to answer in depth.

Before a sound bath:

- Which emotions am I holding onto? How would it feel to let go of them?
- What is making me feel anxious? What am I worried about?
- If I could change one thing about my life with a snap of my fingers, what would I change and why?

After a sound bath:

- If I could choose to be anywhere in the world, where would I go to feel calm?
- If I could do anything I want for a living, what would bring me the most joy and why?
- Think of a fond memory. How did it make you feel at the time? Why is it so special to you?

EMBRACE RITUALS

Whether or not you have children, you will probably be aware that a baby or child will likely have a bedtime routine. You may even have some memories of your own routine from when you were young. The same sequence of events happens each night to signal to a child that bedtime is on its way and let them know what is expected of them.

As adults, we ditch these routines and do not think to include them even if we are struggling with the ability to wind down or relax. Yet our brains love order and predictability and having routines is beneficial. Turning routines into rituals is even better and works exceptionally well when combined with sound healing.

A routine is a series of steps that you may *have* to do – such as wake up, brush your teeth and make your bed. Rituals are also a series of steps, but they do not necessarily *need* to happen. When they do, they have meaning behind them and serve a purpose – such as lighting a candle before your sound bath. The candle in itself is not essential to the enjoyment or nourishment of a sound healing session, but it adds something to the experience.

Rituals also provide the same sense of structure and predictability as a routine. Science shows that this gives a sense of purpose, reduces stress and can help to regulate emotions.[61] A sense of order can be especially important if your life feels particularly chaotic. If you are struggling with anxiety, knowing what to expect as it happens can bring feelings of calm into your day. Plus, since rituals contain meaning, they are more powerful than simply having a routine. Adding a ritualistic element to your sound bath experience will prep your brain, indicate what is about to happen and set the scene for it.

I am a huge fan of rituals. Much of what we discussed in Chapter 6 can be used in a ritualistic way. Rather than just light a candle, turn it into a ritual where you only light a particular candle (or scent) each time you want to meditate with sound. You can choose to light it exactly 20 minutes before your sound experience so that you are fully immersed in the scent. You could decide a particular breathwork exercise is the only one you will use before every sound bath. Or you could always drink a particular flavour of herbal tea after each sonic session while your brain continues to soak up the benefits of resting.

Your rituals can be as small or as big as you want them to be. Some of the healing, stress-beating modalities we have discussed in this chapter can also be incorporated into your rituals. What you include is completely up to you, just as long as it resonates with you, because as with any habit, you have to enjoy it to ensure you stick to it.

Just like attending or listening to a sound meditation, intention is crucial when incorporating a ritual into your practice. There is a big difference between lighting a candle in a hurried, absent-minded way and choosing the same candle and lighting it with intention, watching its flame dance while smelling and appreciating the scent before starting a sound bath. When you do something with intention, you elevate a potential habit – a collective of small, arbitrary steps – into a brain-friendly, stress-busting ritual. It heightens your sense of awareness.

Rituals are also an effective way to incorporate "glimmers" into your life. So often we talk about triggers – situations, people, environments – that can stress us out and make us anxious. Glimmers are the opposite. These are sensory prompts that engage your nervous system to tell you that you are safe and help keep you calm. Having a ritual that you practise each time you have a sound bath experience means that you will come to associate these practices with safety and peace. Then, if you find yourself triggered and can't engage in the full ritual experience of having a sound bath, things will still be okay. Simply taking elements from your ritual – such as smelling the scented candle you always use in your practice – will act as a glimmer to bring you back to a place of regulation.

This is why after a few months of attending regular sound baths, clients may hear the first dong of a bowl or strike of a drum and immediately fall into a deep meditation. Over weeks, they have built up cues to tell them what is going to happen. Their brains are prepped for the sonic experience, enabling them to lean into it immediately and let go even more. They are able to embrace the safety and possibility of where the sound bath may take them more quickly as a result.

Create Your Own Rituals

Once you have built up a regular sound practice, look at making your own rituals so that you can add these into your everyday routine. Grab a pencil and some paper and decide what you may want to do. The key is to have something you can stick to. It is better to have a shorter, firmer ritual that you can ingrain into your brain than something that is too long where you end up missing steps. You can always add more steps afterwards if you want.

Starting small and building it up – i.e. habit stacking – is another good option. Once your weekly sound bath becomes a habit, something that you stick to without much thought or effort, you can add or "stack" another habit to it. Eventually you will have a series of steps that become your personal anti-stress, anti-anxiety ritual.

Think about your environment, too. If you need to let your housemates, partner or kids know that they can't disturb you for the next 40 minutes, do so.

Make sure your ritual is sequential. Do not start with your legs up the wall, then light a candle, then set your intention. Organize what you would like to add to your pre-sound bath ritual in a way that helps it flow best. Set an intention to this time, too; think about how you want to feel at the end of your ritual. And stick to it. To make those connections stronger in the brain, you need repetition, repetition, repetition. Over time, your brain will recognize these cues as your wind-down time before a sound bath.

YOUR SOUND HEALING JOURNEY

Sound healing is a 360-degree way of approaching rest, positivity, change and a fuller, more resonant way of being. So, think of the suggestions in this chapter as little turbo boosters to push you further and welcome even *more* change into your life. You do not have to include these activities with your sound bath practice if you do not wish to. Sound meditation is worthy enough, transformative enough, to stand on its own.

My intention with this book is for it to be as useful as possible, to let you enjoy and develop a true understanding of how melodic, calming sounds created by therapeutic-grade instruments can allow you to reach a place of deep rest. But I also want this book to help you in as many ways as possible. As your relationship with therapeutic sounds deepens, you may find yourself drawn to further exploration. This is where this chapter comes in.

You do not need to do everything I have suggested here. Take what aligns with you and try them when you have the capacity to do so. Incorporate them into your sessions or your day as your exploration with sound grows. Your journey with healing is a personal one. And remember, these suggestions are here to help *you*, to serve *you*, rather than add to your to-do list as yet another task you must complete.

A Final Word

Just as chimes and shakers signify the end of a sound bath, it is now time for me to gently rattle my final words to you. At the end of each group session I lead, I tell my clients that I hope they have enjoyed their time with me. The same applies here – I hope you have enjoyed absorbing what is on these pages, have learnt something valuable and have developed an understanding of why sound is so beneficial, emotionally, mentally and physically.

I have written this book in a way that enables you to come back to it again and again. You may want to try a different technique each time to take your practice to a deeper level. The sound bath audio is for you to enjoy again and again, too. You may find yourself surprised at how your reaction may differ each time you listen to it.

My main hope is that you really sit with what is written in these pages. That developing a better understanding of sound will help you reframe the things in your life that are causing you stress and anxiety. That this understanding will allow you to start healing your body from the noise of the modern world. And that your journey with sound will give you the opportunity to reconnect to your inner voice and wisdom. I hope that you will embrace the concept of resonance to invite more peace and comfort into your life. And that all these changes will allow your nervous system to become calmer and better equipped to deal with any challenges that may occur.

I encourage you to try an in-person session if you have not already done so (maybe even one with me!) and do sample different instruments and varieties of sound offerings that are available. I also hope that you will sit in parks or go for walks by bodies of water and allow yourself to soak up the sounds gifted to us by Mother Nature.

I truly believe that sound can change your life, as it did mine. If I think back to how stressed I was, how dysregulated my nervous system was before I discovered sound healing, I do not recognize my old self. I cannot believe I stayed like that for so long: pushing through, running around the hamster wheel of productivity, ignoring my fundamental needs and thinking rest was not something I required, and could be delayed.

During the pandemic, an ex-colleague listened to one of my online sound offerings and commented on how my voice sounded different to what she remembered. And she was right – my voice has changed. And I hope using my voice through these pages now will help you change yours.

If you would like to connect and share how your journey with sound healing has changed your life, find me @TheSoundTherapist on Instagram.

Online Meditation

Thank you for purchasing *Sound Healing: How to Use Sound to Beat Stress and Anxiety* by Farzana Ali. This book includes a free sound bath meditation. To access this bonus content, scan the below QR code. If the QR code isn't working, please use this link: watkinspublishing.com/sound-healing-3/

Caution: Please be advised that this audio meditation is not suitable for use while driving, operating heavy machinery or any time when peak concentration is required.

Endnotes

1. Thakare, Avinash E et al, "Effect if music tempo on exercise performance and heart rate among young adults", Int J Physiol Pathophysiol Pharmacol, 9(2), Apr. 2017, pp35–39, www.ncbi.nlm.nih.gov/pmc/articles/PMC5435671/

2. Furnham, Adrian and Strbac, Lisa, "Music is a distracting as noise: the differential distraction of background music and noise on the cognitive test performance of introverts and extraverts", Comparative Study, 20;45(3), Feb. 2002, pp203-17, www.pubmed.ncbi.nlm.nih.gov/11964204/

3. Raushch, Vanessa H., Bauch, Eva M. and Bunzeck, Nico, "White noise improves learning by modulating activity in Dopaminergic Midbrain Regions and Right Superior Temporal Sulcus" Journal of Cognitive Neuroscience, 26(7), Jul. 2014, pp1469–80, www.direct.mit.edu/jocn/article-abstract/26/7/1469/28130/White-Noise-Improves-Learning-by-Modulating?redirectedFrom=fulltext

4. Angwin, Anthony J. et al, "White noise enhances new-word learning in healthy adults", Scientific Reports, 7, Oct. 2017, www.nature.com/articles/s41598-017-13383-3

5. Taranto-Montemurro, L et al, "Effect of Background Noise on Sleep Quality" Sleep, 40, Apr. 2017, pp146–7, www.academic.oup.com/sleep/article/40/suppl_1/A146/3781655

6. Kawada, T and Suzuki, S, "Sleep Induction Effects of Steady 60 dB (A) pink noise", Ind Health, 31(1), 1993, pp35–8, www.pubmed.ncbi.nlm.nih.gov/8340228/

7. Kaszubska, Gosia, "Sonic Advance: How sound waves could help regrow bones", RMIT University, Feb. 2022, www.rmit.edu.au/news/all-news/2022/feb/sound-waves-stem-cells

8. Cox, Trevor J., Fazenda, Bruni M. and Greaney, Susan E., "Using scale modelling to assess the prehistoric acoustics of Stonehenge" Journal of Archaeological Science, 122, Oct. 2022, www.sciencedirect.com/science/article/pii/S0305440320301394?via%3Dihub

9. Debertolis, Prof Paolo, Coimbra, Dr Fernando, and Eneix, Linda, "Archaeoacoustic Analysis of Hal Saflieni Hypogeum in Malta" Journal of Anthropology and Archaeology, 3(1), Jun. 2015, pp59-79, www.um.edu.mt/library/oar/bitstream/123456789/16630/1/OA%20Archaeoacoustic%20Analysis%20of%20the%20%C4%A6al%20Saflieni%20Hypogeum%20in%20Malta.pdf

10. Mental Health Foundation, "Stressed Nation: 74% of UK 'overwhelmed or unable to cope' at some point in the past year", UK survey, May 2018, www.mentalhealth.org.uk/about-us/news/stressed-nation-74-uk-overwhelmed-or-unable-cope-some-point-past-year

11. Rovere La, Teresa Maria, Gorini, Alessandra and Schwartz, Peter J., "Stress, the autonomic nervous system, and sudden death", Autonomic Neuroscience, 237, Jan. 2022, www.autonomicneuroscience.com/article/S1566-0702(21)00151-X/fulltext

12. Goh, Joel, Pfeffer, Jeffrey and Zenios, Stefanos, "The Relationship Between Workplace Stressors and Mortality and Health Costs in the United States", Management Science, 62(2) Mar. 2016, pp608–28, www.gsb.stanford.edu/faculty-research/publications/relationship-between-workplace-stressors-mortality-health-costs-united

13. Kraft, Tara L. and Pressman, Sarah D., "Grin and bear it: the influence of manipulated facial expression on the stress response", Psychol Sci., 23(11), Sep. 2012, pp1372–8, www.pubmed.ncbi.nlm.nih.gov/23012270/

14. Nair, Shwetha, Sagar, Mark, Sollers, JJ 3rd, Consedine, Nathan, Broadbent, Elizabeth , "Do slumped and upright postures affect stress responses? A randomized trial", Health Psychology, 34(6), Jun. 2015, pp632–41, www.pubmed.ncbi.nlm.nih.gov/25222091/

15. Murube, Juan, "Hypotheses on the Development of Psychoemotional Tearing", The Ocular Surface, 7(4), Oct. 2009, pp171–175, doi.org/10.1016/S1542-0124(12)70184-2

16. Fisher, Jen, "Workplace Burnout Survey", 2015, www2.deloitte.com/us/en/pages/about-deloitte/articles/burnout-survey.html

17. Hatfield, St eve, Fisher, Jen and Silverglate, Paul H., "The C-suite's role in well-being", Deloitte Insights, 22(7), Jun. 2022, www2.deloitte.com/us/en/insights/topics/leadership/employee-wellness-in-the-corporate-workplace.html

18. Pencavel, John, "The Productivity of Working Hours", Discussion Paper Series, (8129), Apr. 2014, docs.iza.org/dp8129.pdf

19. Dalton-Smith, Saundra, Sacred Rest: Recover Your Life, Renew Your Energy, Restore Your Sanity, Faithwords, 2018

20. Vargas-Caballero, Mariana et al, "Vagus Nerve Stimulation as a potential Therapy in Early Alzheimer's Disease: A Review", Front Hum Neuroscience, 16, Apr 2022, www.ncbi.nlm.nih.gov/pmc/articles/PMC9098960/

21. Marsal, Prof Sara, "Non-invasive vagus nerve stimulation for rheumatoid arthrist: a proof-of-concept study", The Lancet Rheumatology, 3(4), Feb. 2021, pp262–269, www.thelancet.com/journals/lanrhe/article/PIIS2665-9913(20)30425-2/

22. Dyer, Wayne W, Change Your Thoughts, Change Your Life: Living the Wisdom of the Tao, Hay House, 2006

23. Nin, Anais, www.goodreads.com/quotes/5030-we-don-t-see-things-as-they-are-we-see-them

24. Sinek, Simon, "How to stop holding yourself back", 17(7), 2021, www.facebook.com/simonsinek/videos/how-to-stop-holding-yourself-back/950325835533422

25. van der Kolk, Bessel, The Body Keeps the Score: Brain, Mind, and Body in the Healing of Trauma, Penguin, 2005

26. Breit, Sigridm et al, "Vagus Nerve as Modulator of the Brain-gut Axis in Psychiatric and Inflammatory Disorders", Front Psychiatry, 9, Mar. 2018, www.ncbi.nlm.nih.gov/pmc/articles/PMC5859128/

27. Goldsby, Tamara L. et al, "Effects of Singing Bowl Sound Meditation on Mood, Tension, and Well-being: an Observational Study", J Evid based Complementary Alter Med, 22(03), Jul. 2017, pp401-406 www.ncbi.nlm.nih.gov/pmc/articles/PMC5871151/

28. Herrando, Carolina, and Constantinides, Efthymios, "Emotional Contagion: A brief overview and future directions", Front Psychology, 12, Jul. 2021, www.ncbi.nlm.nih.gov/pmc/articles/PMC8322226/

29. Cudjoe, Thomas K. M. et al, "Getting under the skin: Social isolation and biological marker in the National Health and Aging Trends Study", J Am Geriatr Soc., 70(2), Feb 2022, pp408-414, www.pubmed.ncbi.nlm.nih.gov/34698366/

30. Van Bogart, Karina et al, "The Association Between Loneliness and Inflammation: Findings From an Older Adult sample", Front Behav Neurosci, 15, 2021 www.ncbi.nlm.nih.gov/pmc/articles/PMC8787084/

31. Jimenez, Marcia P. et al, "Associations between Nature Exposure and Health: a Review of the Evidence", Int J Environ Res Public Health, 18(9), Apr. 2021, www.ncbi.nlm.nih.gov/pmc/articles/PMC8125471/

32. Schertz, K. E. and Berman, M. G., "Understanding Nature and Its Cognitive Benefits", Current Directions in Psychological Science, 28(5), Jun. 2019, pp496–502, journals.sagepub.com/doi/10.1177/0963721419854100

33. Zelenski, John M., Dopko, Raelyne L., and Capaldi, Colin A., "Cooperation is in our nature: Nature exposure may promote cooperative and environmentally sustainable behaviour", Journal of Environmental Psychology, 42, Jun. 2015, pp24–31, www.sciencedirect.com/science/article/pii/S0272494415000195?via%3Dihub

34. Pritchard, Alision et al, "The Relationship Between Nature Connectedness and Eudaimonic Well-being: A Meta-analysis", Journal of Happiness Studies, 21, 2020, pp1145–67, link.springer.com/article/10.1007/s10902-019-00118-6

35. Capaldi, Colin A. et al, "Flourishing in nature: A review of the benefits of connecting with nature and its application as a wellbeing intervention", International journal of Wellbeing, 5(4), Dec 2015, www.internationaljournalofwellbeing.org/index.php/ijow/article/view/449

36. Park, Bum Jin et al, "The physiological effects of Shinrin-yoku (taking in the forest atmosphere or forest bathing): evidence from field experiments in 24 forests across Japan", Environ Health Prev Med, 15, 2010, pp18–26, www.drperlmutter.com/study/the-physiological-effects-of-shinrin-yoku-taking-in-the-forest-atmosphere-or-forest-bathing-evidence-from-field-experiments-in-24-forests-across-japan/

37. Li, Q. and Kobayashi, M. et al, "Effect of Phytoncide from trees on human natural killer cell function", International Journal of Immunopathology and Pharmacology, 22(4), Oct 2009, pp951–959, www.researchgate.net/publication/41027513_Effect_of_Phytoncide_from_Trees_on_Human_Natural_Killer_Cell_Function#:~:text=Phytoncide%20exposure%20significantly%20increased%20NK,adrenaline%20and%20noradrenaline%20in%20urine.

38. University Communications and Marketing, www.calpolynews.calpoly.edu/news_releases/2020/december/birdsongs

39. Basner, Mathias et al, "Auditory and non-auditory effects of noise on health", Lancet, 383(9925), 2014, pp1325-1332 wwwncbi.nlm.nih.gov/pmc/articles/PMC3988259/

40. Ulrich, Roger S. et al, "Stress recovery during exposure to natural and urban environments", Journal of Environmental Psychology, 11(3),1991, pp201–30, www.psycnet.apa.org/record/1992-11140-001

41. Van Hedger, Stephen C. et al, "Of cricket chirps and car horns: The effect of nature sounds on cognitive performance", Psychonomic Bulletin & Review, 26, 2019, pp522–30, www.link.springer.com/article/10.3758/s13423-018-1539-1

42. Francis, Clinton D et al, " Acoustic environment matter: Synergistic benefits to humans and ecological communities", J Environ Manage, 203, Dec.2017, pp245–54, www.pubmed.ncbi.nlm.nih.gov/28783021/

43. Oschman, James L, Chevalier, Gaétan, and Brown, Richard, " The effects of grounding (earthing) on inflammation, the immune response, wound healing, and prevention and treatment of chronic inflammatory and autoimmune diseases.", J Inflamm Res, 8, Mar. 2015, pp83–96, www.ncbi.nlm.nih.gov/pmc/articles/PMC4378297/

44. Chevalier, Gaétan et al, "Earthing: Health Implications of Reconnecting the Human Body to the Earth's Surface Electrons", J Environ Public Health, 2012, Jan. 2012, www.ncbi.nlm.nih.gov/pmc/articles/PMC3265077/

45. Zhou, Junhong et al, "Pink noise: effect on complexity synchronization of brain activity and sleep consolidation", J Theor Biol, 306, Aug. 2012, pp68–72, www.pubmed.ncbi.nlm.nih.gov/22726808/

46. Papalambros, Nelly A., and Santostasi, Giovanni, and Malkani, Rineil G., et al, "Acoustic Enhancement of Sleep Slow Oscillations and Concomitant Memory Improvement in Older Adults.", Frontiers in Human Neuroscience, 11, Mar. 2017, www.frontiersin.org/articles/10.3389/fnhum.2017.00109/full

47. Qureshi, Andan et al, "Cat ownership and the Risk of Fatal Cardiovascular Diseases. Results from the Second National Health and Nutrition Examination Study Mortality Follow-up Study", J Vasc Interv Neurol, 2(1), Jan. 2009, pp132–35, www.ncbi.nlm.nih.gov/pmc/articles/PMC3317329/

48. Fernando, Gunasekara et al, "Adjunctive Effects of a Short Session of Music on Pain, Low-mood and Anxiety Modulation among Cancer Patients- A Randomized Crossover Clinical Trial", Indian J Palliat Care, 25(3), July-September 2019,pp367–73, www.ncbi.nlm.nih.gov/pmc/articles/PMC6659521/

49. "Classical Music Eases Arthritis Symptoms", Arthritis Today Magazine, June 2015, blog.arthritis.org/living-with-arthritis/classical-music-arthritis/

50. Nilsson, Ulrica, "Soothing music can increase oxytocin levels during bed rest after open-heart surgery: a randomized control trial", Journal of Clinical Nursing, 18(15), 2009, pp2153–61, www.onlinelibrary.wiley.com/doi/full/10.1111/j.1365-2702.2008.02718.x

51. Colver, Mitchell C. and El-Alayli, Amani, "Getting aesthetic chills from music: The connection between openness to experience and frisson", Psychology of Music, 44(3), 2015, pp413–27, www.journals.sagepub.com/doi/abs/10.1177/0305735615572358

52. Swee, Genevieve and Schirmer, Annett, "On the importance of being vocal: saying 'ow' improves pain tolerance", J Pain, 16(4), Apr. 2015, pp326–34 www.pubmed.ncbi.nlm.nih.gov/25622894/

53. Baker, Lois, "Lung Function May Predict Long Life or Early Death", University at Buffalo Newsletter, Sep. 2000, www.buffalo.edu/news/releases/2000/09/4857.html

54. Kreutz, Gunter et al, "Effects of choir singing or listening on secretory immunoglobulin A, cortisol, and emotional state", J Behav Med., 27(6), Dec 2004, pp623–35, www.pubmed.ncbi.nlm.nih.gov/15669447/

55. Levine, Arlene Bradley, Punihaole, David, and Levine, Barry T., "Characterization of the Role of Nitric Oxide and Its Clinical Applications", Cardiology, 122(1), 2012, pp55–68, www.karger.com/crd/article-abstract/122/1/55/76503/

56. Togo, Takashi, Katsuse Omi, and Iseki Eizo, "Nitric oxide pathways in Alzheimer's disease and other neurodegenerative dementias", Neurol Res, 26(5), 2004, pp563–6, pubmed.ncbi.nlm.nih.gov/15265275/

57. Gao, Junling et al, "Repetitive Religious Chanting Modulates the Late-Stage Brain Response to Fear- and Stress-Provoking Pictures", Front Psychol, 7, 2016,, pp2055, www.ncbi.nlm.nih.gov/pmc/articles/PMC5223166/

58. Dharma Singh, Khalsa et al, "Cerebral blood flow changes during chanting meditation", Nucl Med Commun, 30(12), 2009, pp956–61, www.pubmed.ncbi.nlm.nih.gov/19773673/

59. Perry, Gemma, Polito, Vince, and Forde Thompson, William , "Chanting Meditation Improves Mood and Social Cohesion", Proceedings of the 14th International Conference on Music Perception and Cognition, 14,Jul 2016, pp324-327 www.researchgate.net/publication/319851087_Chanting_Meditation_Improves_Mood_and_Social_Cohesion

60. Norman-Haignere, Sam V. et al, "A neural population selective for song in human auditory cortex", Cell Press Journal, 32(7), Apr. 2022, pp1470–84, www.cell.com/current-biology/fulltext/S0960-9822(22)00131-2

61. Poerio, Giulia Lara et al, "More than a feeling: Autonomous sensory meridian response (ASMR) is characterized by reliable changes in affect and physiology", PLoS One, 13(6), Jun 2018, www.ncbi.nlm.nih.gov/pmc/articles/PMC6010208/

62. Lin, Frank R., Metter, Jeffrey E., and O'Brien, Richard J., "Hearing Loss and Incident Dementia", Arch Neurol, 68(2), Feb. 2011, pp214–20, www.jamanetwork.com/journals/jamaneurology/fullarticle/802291

63. Cuddy, Amy J.C., Wilmuth, Caroline A., and Carney, Dana R., "The Benefit of Power Posing Before a High-Stakes Social Evaluation", Harvard Business School Working Paper, 2012, pp13-27, www.dash.harvard.edu/bitstream/handle/1/9547823/13-027.pdf?sequence=1

64. Kraft, Tara, "Grin and Bear it! Smiling Facilities Stress Recovery", Psychological Science, 23, Jul 2012, www.psychologicalscience.org/news/releases/smiling-facilitates-stress-recovery.html

65. Zhao, Ziyi et al, "Effects of mouth breathing on facial skeletal development in children: a sustematic review and meta-analysis", BMC Oral Health, 21(108), Mar. 2021, www.ncbi.nlm.nih.gov/pmc/articles/PMC7944632/

66. Jefferson, Yosh, "Mouth breathing: adverse effects on facial growth, health, academics and behavior", Gen Dent, 58(1), Jan.–Feb. 2010, pp18–25, quiz pp26-7, pp79-80 www.pubmed.ncbi.nlm.nih.gov/20129889/

67. Bordoni, Bruno et al, "The Anatomical Relationships of the Tongue with the Body System", Cureus, 10(12), Dec. 2018, www.ncbi.nlm.nih.gov/pmc/articles/PMC6390887/

68. Ramirez, Jan-Marino, "The Integrative Role of the Sigh in Psychology, Physiology, Pathology and Neurobiology", Prog Brain Res, 209, 2014, pp91–129, www.ncbi.nlm.nih.gov/pmc/articles/PMC4427060/

69. Naik, G. Sunil, Gaur, G.S., and Pal, G.K., "Effect of Modified Slow Breathing Exercise on Perceived Stress and Basal Cardiovascular Parameters", Int J Yoga, 11(1), Jan.–Apr. 2018, pp53–8, www.ncbi.nlm.nih.gov/pmc/articles/PMC5769199/

70. Weitzberg, Eddie and Lundberg, Jon O. N., "Humming Greatly Increases Nasal Nitric Oxide", American Journal of Respiratory and Critical Care Medicine, 166(2), 2022, www.atsjournals.org/doi/full/10.1164/rccm.200202-138BC

71. Emmons, Robert A and McCullough, Michael E., "Counting blessings versus burdens; an experimental investigation of gratitude and subjective well-being in daily life", J Pers Soc Psychol, 84(2), Feb. 2003, pp377–89, www.pubmed.ncbi.nlm.nih.gov/12585811/

72. Allen, Summer, "The Science of Gratitude", Greater Good Science Center, May 2018, www.ggsc.berkeley.edu/images/uploads/GGSC-JTF_White_Paper-Gratitude-FINAL.pdf

73. Hobson, Nicholas M., and Bonk, Devin, and Inzlicht, Michael, "Rituals decrease the neural response to performance failure", Peer J, 5, May 2017,

www.ncbi.nlm.nih.gov/pmc/articles/PMC5452956/

Further Reading

The Singing Bowl Book by Joseph Fienstien, Independently Published, 2018

Cosmic Octave: Origin of Harmony by Hans Cousto, LifeRhythm, 2001

Musicophilia: Tales of Music and the Brain by Oliver Sacks, Picador, 2011

This Is Your Brain on Music: The Science of a Human Obsession by Daniel Levitin, Penguin, 2019

"'Tibet Chic': Myth, Marketing, Spirituality and Politics in Musical Representations of Tibet in the United States" by Darinda J. Congdon BM, Baylor, 1997 MA, University of Pittsburgh, 2002. d-scholarship.pitt.edu/8589/1/DCongdon_ETDAug2007.pdf

Glossary

ASC – An Altered State of Consciousness. A trance-like mental state.

Auditory cortex – Where sound is perceived in the brain.

Archaeoacoustics – A field of archaeology that studies sound and acoustics at archaeological sites across different cultures.

Brainwaves – The electrical signature of brain activity as neurons communicate with each other.

Binaural beats – An auditory illusion where two different tones are played in each ear, creating a third, new frequency tone in the brain.

Chord – A group of notes played together.

Entrainment/sympathetic resonance – The synchronization of something to an external rhythm.

Frequency – The rate at which sound waves vibrate.

Harmony – The pleasing combination of two or more notes played at the same time.

Hamonic – When two or more notes accompany a lower, fundamental note.

Hertz (Hz) – Unit of measurement for frequency. Both sound waves and human brain waves are measured in Hertz.

Interval – The musical distance between two notes. Different intervals can elicit different emotional responses.

Parasympathetic nervous system – The "rest and digest" part of the autonomic nervous system. The opposite of the sympathetic (aka fight-or-flight) nervous system.

Pitch – How high or low a note is.

Rhythm – The pattern of beats.

Sound bath – The name given to any meditative session that uses sound to induce a state of mental, emotional and physical rest.

Sound therapy – The use of therapeutic techniques combined with sound healing instruments.

Sound wave – Movement of vibrating energy travelling through a medium (such as air, water or solid matter) as it moves away from the source of the sound.

Vagus nerve – One of the 12 cranial nerves of the body. It controls a variety of essential bodily functions such as heart rate, breath rate and digestion. It makes up 75 per cent of your parasympathetic nervous system.

Acknowledgements

No journey happens in isolation. Taking a different path and overhauling one's life and career is not easy and can only be achieved with the guidance, love and encouragement of others. Community is at the heart of all success – no matter how you define that. Just like sound waves, one bit of kindness or support sends out ripples.

My thanks go to my community, my circle, my family and friends who have been pivotal in empowering me to realize every dream, least of all this dream of writing a book.

To my sister – without your support, literally nothing would have been possible. From gifting me instruments, to lending me your house and propping up my finances when I was studying, and even buying me new furniture after "The Flood" (page 107), every single act of kindness from you has enabled me to embark on a new life course.

To my friends, who have helped and supported me. From designing logos to giving me business tips, for pitching my services to your clients and bosses; to mentioning my name in rooms where I have not been present but opportunities have been. I am forever thankful to each and every one of you. I am also so awestruck by people's generosity, for helping and supporting me – even if they have only met me once, or not at all – for opening doors, for putting me forward. I am so grateful to be around people who celebrate my wins as their own.

To my father, for raising me in a way that led me to believe that anything, any goal, was possible.

To my mother, for your prayers.

To my wonderful clients for trusting me, for sharing your experiences with me, for coming back each week or as you need, and for making all the learning worth it.

To my editor Lucy – thank you for sliding into my DMs and giving me this unbelievable opportunity and for believing in me.

To my copyeditor Victoria – for your patience.

Finally, to the Most High, the Provider, the Healer, the Creator of the Universe, for the direction and protection that has led me here. Alhamdulillah, I am thankful for all my blessings.

About the Author

Farzana Ali, also known as The Sound Therapist to her clients, is a practitioner-level sound therapist, meditation teacher and wellness expert. She is also a former health and lifestyle editor. Taking a science-based, trauma-informed view of sound meditation, she believes in the full transformative power of sound.

Her work has already seen her collaborate with major health, wellness and beauty brands and hotels around the world, allowing her to continually introduce sound healing to a wider audience. Her TV appearances include ITV's *This Morning* and Garraway's *Good Stuff* and she has been featured by *British Vogue*, *Harper's Bazaar*, *Women's Health*, *Elle*, *Stylist*, as well as UK national newspapers including *The Guardian*, *Daily Express* and *The Metro,* to name a few.

To connect with Farzana head to:
Instagram: @thesoundtherapist
Website: www.thesoundtherapist.com
Newsletter: Farzanaali.substack.com
Youtube: @FarzanaTheSoundTherapist